Abundant Earth

Abundant Earth

Toward an Ecological Civilization

EILEEN CRIST

The University of Chicago Press   Chicago and London

The University of Chicago Press, Chicago 60637
The University of Chicago Press, Ltd., London
© 2019 by The University of Chicago
Published 2019
Printed in the United States of America

28 27 26 25 24 23 22 21 20 19   1 2 3 4 5

ISBN-13: 978-0-226-59677-8 (cloth)
ISBN-13: 978-0-226-59680-8 (paper)
ISBN-13: 978-0-226-59694-5 (e-book)
DOI: https://doi.org/10.7208/chicago/9780226596945.001.0001

Library of Congress Cataloging-in-Publication Data

Names: Crist, Eileen, 1961– author.
Title: Abundant Earth : toward an ecological civilization /
    Eileen Crist.
Description: Chicago ; London : The University of Chicago Press,
    2019. | Includes bibliographical references and index.
Identifiers: LCCN 2018017666 | ISBN 9780226596778 (cloth :
    alk. paper) | ISBN 9780226596808 (pbk. : alk. paper) |
    ISBN 9780226596945 (e-book)
Subjects: LCSH: Biodiversity conservation. | Human-animal
    relationships. | Human-plant relationships.
Classification: LCC QH75 .C735 2018 | DDC 333.95/16—dc23
LC record available at https://lccn.loc.gov/2018017666

♾ This paper meets the requirements of ANSI/NISO Z39.48–1992
(Permanence of Paper).

Contents

Acknowledgments

I thank my good friends and colleagues, from whom I have learned so much and with whom I have enjoyed instructive conversations over the years: David Abram, Marc Bekoff, Joe Bish, Tom Butler, Phil Cafaro, Matthew Calarco, Martha Campbell, Bryce Carter, Lauren Cooper, Chris Cox, Patrick Curry, John Davis, Marcia Davitt, Dominick DellaSalla, Adam Dickerson, Eric Dinerstein, Anne Ehrlich, Paul Ehrlich, Robert Engelman, Tim Filbert, Dave Foreman, Joe Gray, Donna Haraway, Brian Henning, Sandy Irvine, Wes Jackson, Dale Jamieson, Derrick Jensen, David Johns, Paul Keeling, Christopher Ketcham, Lierre Keith, David Kidner, Helen Kopnina, Lisi Krall, Jenn Lawrence, Harvey Locke, Damien Mander, Jerry Mander, Richard Manning, Douglas McCauley, Bill McKibben, Stephanie Mills, George Monbiot, Camilo Mora, Roderick Nash, David Nibert, Richie Nimmo, Reed Noss, David Orr, Stuart Pimm, Luke Philip Plotica, H. Bruce Rinker, William Ripple, Callum Roberts, Holmes Rolston, Deborah Bird Rose, William Ryerson, Ken Smail, Michael Soulé, Gary Steiner, Alfred Tauber, Bron Taylor, Doug Tompkins, Kristine McDivitt Tompkins, Jack Turner, Will Tuttle, Thom van Dooren, Sacha Vignieri, John Waldman, Haydn Washington, Don Weeden, Tony Weis, Alan Weisman, Ian White, Bryn Whiteley, Terry Tempest Williams, E. O. Wilson, and George Wuerthner. I also want to acknowledge a huge debt to my undergraduate and graduate school teachers, most especially Mark Gould, Michael Lynch, Jeff Coulter, Lynn Margulis, and Frederick Wasserman.

I wish also to express my gratitude to all my colleagues

in the Department of Science, Technology, and Society at Virginia Tech—one could not wish for more supportive and warmhearted people to work with. Also a big thank-you goes out to the Department of Environmental Studies of New York University for hosting me as a visiting scholar in 2014–15. I am indebted to funding from the Foundation for Deep Ecology and the Don Weeden Foundation, which made my research year in New York City possible. I would like to thank my University of Chicago editors, Christie Henry, Miranda Martin, and Mark Reschke, for their enthusiasm, professionalism, diligence, and support. I also thank Pascal Pocheron of Polity Press for generous feedback on an early draft of my book proposal.

I am forever grateful to the guiding lights in my life, Dr. Robert E. Svoboda, Tenzin Wangyal Rinpoche, Alejandro Chaoul, Marcy Vaugn, Dharma Mittra, Krishna Das, Sharon Gannon, and David Life. I am also deeply grateful to my parents, Robert Crist and Despina Lala-Crist, who raised me on nature and poetry, and to my beloved brother Ray Crist who is an original thinker and practitioner in his own right. I extend my heartfelt gratitude to my second family, Cynthia and Dennis Patzig, Joanna Patzig, Zach Patzig, and Alice Lee, for their love and support. Last, I thank my husband, Rob Patzig, for being in my life, for his compassion, strength, and illumination. I dedicate this work to him.

Introduction

Cosmologist Brian Swimme relates a personal story that captures the conundrum of life's crisis in our time. After hearing the announcement of a meeting of leading life scientists that humanity's impact is heading the biosphere toward a mass extinction, he went to bed that night deeply disturbed. First thing in the morning, he reached for the *New York Times* to see how this earth-shattering news was reported in the media. Page after page there was nothing. Finally, on page 26 he found a terse report of the announcement. Swimme's shock at the media's underwhelming reception was spot on: The *New York Times* found twenty-five pages of more important reporting than the news of a human-driven mass extinction on the horizon.

The Earth has indeed come upon hard times. With an estimated extinction rate one thousand times higher than the natural rate of extinction, a mass extinction event looms. Species and subspecies are disappearing, most before we get to meet them. Huge declines in populations of wild animals and plants, as well as the destruction of wholesale ecologies, are occurring across the board. Phenomena of biological abundance, like animal migrations and wildlife spectacles, are disappearing. Two recent findings speak volumes. In the last fifty years, more than half the Earth's wild animals disappeared. In the last forty years, 10 percent of Earth's already contracted wilderness was destroyed. Without a profound shift in humanity's historical course, the biosphere will soon become completely dominated by human beings, domestic species, built structures, industrial infrastructures, and a cadre of globally recurring species able to survive amidst such conditions.

In the wake of countless scientific publications in recent decades, what is driving the steep decline of life's richness is clear: humanity's expansionism of growing economies, escalating global trade, climbing population numbers, sprawling infrastructures, and spreading destructive technologies. Despite knowledge of what is transpiring and where the biosphere is headed, the takeover of the natural world to make way for food production, materials and energy extraction, commodity generation, infrastructural gridding, and all manner of development continues unabated.

Abundant Earth focuses on the demolition of life's variety, complexity, and plenitude, with the aim of unmasking the widely shared belief system of human superiority and entitlement that undergirds humanity's destructive expansionism. Key quandaries of the book's narrative echo Swimme's incredulity: Why is the collapse of biological diversity sidelined in mainstream culture? And relatedly: Since it is well understood that human expansionism is causing life's crisis, why is humanity not taking steps to halt its expansionism? Addressing these questions through scholarly analysis and critique is necessary, in my view, for it is comparable to removing the veils that make the annihilation of life's richness invisible and mostly ignored. Importantly, such analysis and critique sets the stage for elaborating the possibility of an altogether different relationship between humanity and the Earth: the choice to scale down and pull back the human presence and to pursue the creation of a global ecological civilization within the planet's full house of life.

Part 1 maps the collapse of biodiversity occurring in our time, documenting its direct causes and underlying drivers. This mapping is intended to convey the systemic scope of life's destruction. Earth's variety of life-forms, diversity of unique natures, abundance of wild plants and animals, expansiveness of untamed places, and complexity of ecologies and phenomena are dwindling and vanishing. Without a profound shift to match the catastrophe's magnitude, the ecological crisis is heralding the entrenchment of the domination of nature and the total repurposing of Earth as humanity's resource colony.

Simply clarifying the direct and ultimate causes of life's crisis does not get us any closer to understanding why humanity is doing vanishingly little to address it. On the contrary, barefaced inaction—despite established knowledge of the causes of life's devastation—begs that question. I argue that the answer lies in the reigning worldview of human supremacy (or anthropocentrism) that stands as an intractable obstacle to the historical shift required, because it both normalizes and promotes ongoing human expansionism. Human supremacy is the collective, lived belief system that humans are superior to all other life-forms and entitled to use them and their places of livelihood.

This worldview makes humanity's planetary sovereignty appear as a world order that is indisputably given. Sociocultural conditioning into the precepts of human distinction and prerogative renders the very notion of substantially scaling down and pulling back humanity's sprawl almost unthinkable from a mainstream perspective. Indeed, the human-supremacist worldview stifles receptivity to the tack of contracting the human project so as to sustain the plenum of life on Earth. The approach of humanity's scaling down and pulling back is systematically ignored, or at best marginalized, in the dominant culture and its policy frameworks, which vaunt the specialness, privilege, and rightful perquisite of the human over the face of the Earth.

In lieu of recognizing that ceding human dominance is the only resolution for preserving and restoring life's richness, mainstream venues regularly hype technological and managerial avenues for addressing arising challenges—avenues that diligently avoid questioning, let alone confronting, humanity's colonization of the biosphere. Techno-managerialism—the go-to framework of policy circles, nation-states, corporate entities, research centers, and most universities—aspires to sustain the status quo of Earth as the Planet of the Humans, while striving to serially mitigate or fix any civilization-endangering catastrophes brought on by that status quo.

More surprisingly, the human-supremacist worldview is underchallenged in the environmental domain as well, which has as its mission to clarify human-nonhuman unequal power relations and to offer alternative pathways forward. The question of why environmental thought has largely desisted from opposing anthropocentrism, standing up for nature's freedom, and agitating for the end of human expansionism deserves attention. Part 2 of this work investigates certain "discursive knots" that impede the environmental movement from becoming a genuinely game-changing force—a force that will resist the life-destroying human-supremacist worldview and inspire humanity to move in the direction of a life-affirming and life-abundant vision to live by.

The metaphor of discursive knots is repurposed from Buckminster Fuller's definition of a knot as "an interfering pattern." We are all familiar with the fact that a knot becomes more and more difficult to undo as additional knots are piled on top of the original one. Analogously, discursive knots are oft rehearsed patterns of reasoning about the global situation that interfere with the flow of imagination and action in an alternative direction: namely, of contesting what we might call "the trends of more"—burgeoning economies and global trade, a growing human population, rising livestock numbers, multiplying extractive enterprises, and invading infrastructures—instead of regarding them as unchangeable variables around which techno-managerial adjustments must be applied and a mood of resilience (in the wake of consequences) must be boostrapped.

I tackle the interfering patterns of three discursive knots: a widespread proclivity to view the human impact as natural; a fashionable trend to concede wilderness as defunct reality and bankrupt idea; and a standard acclaim of human expansionism as salutary for bringing more and more freedoms to increasing numbers of people. These are exceedingly prevalent ideas in our time with correspondingly formidable repercussions.

Naturalizing the human impact is a common inclination to attribute the human onslaught on the natural world to peculiarities of our species' makeup—to "human nature," in short. Naturalizing humanity's onslaught is a knot of Gordian proportions, because so many people adhere to the view (whether vaguely or rigorously formulated) that human nature is essentially the culprit behind the ecological crisis. The circulating sound bite "we have met the enemy and it is us" captures the essence of this conviction. As a consequence, critical thinking about the ways humans are socioculturally programmed into a supremacist worldview—one that effectively construes the natural world as beneath and for the human—is thoroughly obstructed.

Naturalizing the human impact also blocks the awareness from dawning that ending human domination of and dominance within the biosphere is the only pathway to resolving the ecological crisis. That direction of thought and action is disabled by an inexorable implication of naturalizing human impact: if humanity's onslaught is a direct upshot of "human nature," then the event of the onslaught itself is a straightforward extension and expression of the natural order; while dangers galore may trail the human impact, there is nothing existentially or ethically amiss about it. Such a perspective encourages acceptance of (or resignation to) the status quo, and tacitly bolsters the belief that the best we can do is to pursue technological transitions, effect damage control, clean up egregious side effects, muddle through serial challenges or crises as they arise, and just plain hope for the best.

Naturalizing the human impact does not move people in the direction of fundamentally changing how we inhabit Earth.

A recent trend to dispense with wilderness as an empirically and ideologically flawed notion is the second discursive knot examined. This belief has become so diffuse and banal that many people dismiss "wilderness" as a shibboleth that no longer merits our critical attention. To be sure, humanity's impact is profound and pervasive, having left no place on Earth untouched—from the stratosphere to the Mariana trench and from pole to pole. Yet dispensing with wilderness on the basis of this is like throwing out the baby with the bathwater. For to discard the conceptual vessel of wilderness—as referent to the natural world and as an idea in the human imagination—is to forsake nature's original blueprint, which lucidly reflects back the impoverishment that limitless human expansionism has produced. To forsake nature's original blueprint of prodigious creativity and abundance supports the allowance of its banishment in both reality and memory.

I argue that relinquishing wilderness, as certain strands of environmental thought have encouraged, has done an extraordinary disservice to the Earth and to human possibilities of being in the world, by undermining a spirited defense of nature's autonomy and of wild beings, who are, right now, experiencing extreme suffering and dislocations as well as untimely death and extinction. Setting aside facile verdicts, such as "wilderness is gone" or "wilderness is just a sociocultural construct," is imperative for seeking fresh insight into the meaning of nature's freedom and for inspiring a social movement that will defend wild nature's qualities of unrivaled creativity and abundance.

Yet even as nature's freedom is screaming for defense, the offensive of humanity's expansionism is piously touted as promoting human freedoms—freedoms secured by enlarging the means and scope of human mobility, spreading modern conveniences, multiplying the glut of commodities and foods to choose from, enabling horizon-expanding experiences in far-flung places and exotic lands, and enlarging the possibilities of virtual interconnectivity. The third discursive knot I unravel is the framing of expansionism as delivering modern freedoms to increasingly more people. This frame is the premier ideological boost of the explosive growth in mechanized mobility, commodities markets, global trade, communications technologies, and industrial infrastructures—growth effected directly at the expense of the natural world and built out of its demolition.

Indeed, in a world of billions, growing in both numbers and affluence, spreading modern freedoms is premised on extinguishing the freedoms of the nonhuman realm. Earth's web of life is undone and downgraded to facilitate the unrestricted human experience of movement, access, use, acquisition, consumption, travel, entertainment, and connectivity. These ostensible privileges necessitate eradicating nonhuman freedoms and destroying free nature (wilderness). The grating incoherence of securing human freedoms by means of exterminating nonhuman freedoms precisely motivates the silence enveloping the collapse of biological diversity and the imminent mass extinction event in the mainstream culture. This oversight is not incidental: the implosion of life's richness has to remain obscure in public consciousness, as it is a direct upshot of the freedoms that people (are incited to) value and seek. Avoiding clarity about this Faustian, defining deal of our time is exceptionally serviceable, for most reasoning human beings are well aware that no authentic freedom for oneself can be founded on taking away freedom from others. Unraveling the discursive knot that conflates human expansionism with the spread of freedoms calls us to break the silence that conspires to conceal the reign of death those "freedoms" demand.

Deconstructing the muddled and violent project of founding human freedom on the destruction, constriction, exploitation, and enslavement of the larger community of life (wild and domestic) opens a view to another possibility: that of thinking deeply about what human freedoms within the biosphere will look like once we embrace a broadened ideal of freedom for all Earth's inhabitants—and indeed for the Earth itself. It does not follow from this argument that modern freedoms must be completely relinquished, nor that human beings must don Franciscan robes of austerity. It follows that humanity must welcome limitations for the sake of a higher vision and practice of freedom. Such enlightened intent grounding the embrace of limitations reveals that scaling down and pulling back the human presence are neither a sacrifice nor a contraction of human potential, but on the contrary harbor the blossoming of human virtue and the coflourishing of all earthly life.

Part 3, "Scaling Down and Pulling Back," explores a beautiful way forward that will enable the nonhuman and human realms to thrive together. Scaling down means drastically reducing consumption and waste, which, along with other needed actions, mandates lowering the global population, deindustrializing food production, localizing economies, and greatly reducing global trade. Pulling back refers to the project of restoring, reconnecting, and rewilding vast portions of land and ocean so as to enable life's plenum to surge. I argue that scaling down and pulling back configure the conjoined strategy for moving toward a global ecological civilization within a biodiverse planet.

To drive home the imperative of curtailing the human project, while simultaneously setting Earth free to express its living arts, it is useful to understand the dystopian world toward which the biosphere is headed should "the trends of more" be left to unfold unchecked. The dystopia at the doorstep is a humanized world dominated by industrial agriculture and aquaculture, with threadbare biodiversity and without "a blank spot on the map."[1] On such a "human eminent domain" planet, every last resource of materials and energy that can be gouged and flushed out—from the most forbidding places by the most extreme technologies—will be extracted by corporations and nation-states. Humanity is rapidly plunging the biosphere toward totalitarian rule, managed as a food-production plantation, engineered for harnessing energy, scoured for materials extraction, crisscrossed by all manner of infrastructures, overrun by billions of automobiles, dominated by bloated commodity markets, within which the chief human identity of user will become hardened. Totalitarian rule will generate a world that will require "securitizing" at every level, for the human condition is bound to be haunted (as it in fact already is) by actual, probable, and possible crises of unprecedented and unpredictable proportions.

Humanity has the choice to veer away from the supremacist historical project of colonizing the Earth (and to thus preempt the sundry grim repercussions of that project), toward a way of life that preserves Earth's beautiful cadre of abundant life, sustains wild nature's rambunctious freedom, and keeps life's flame blazing in the cosmos. Espousing that choice means nothing other and nothing less than shrinking the human presence. I submit that the Ariadne's thread of this historical redirection is overhauling the most ecologically destructive enterprise on Earth: the industrial food system (meaning industrial production, manufacturing, and trade). The industrial food system is implicated in virtually every human-driven global scourge: the collapse of marine life, the extinction crisis, the devastation of big carnivores and herbivores, the freshwater biodiversity crisis, rapid climate change, rampant fertilizer and pesticide runoff, plastic ocean pollution, honeybee colony collapse disorder, the bushmeat crisis, and the destruction of ecologies from jewel-small wetlands to rain forest and grassland biomes. The industrial food system—serving an enormous and growing human population increasingly connected as a global consumer society—is hitched to every major affliction of the planet.

The food system of an ecological civilization will eschew chemical pesticides and fertilizers, dismantle large-scale monoculture production, shrink animal agriculture, abolish industrial fishing, interface creatively with wild nature, and be designed primarily to support human beings locally and regionally. Ecologically sound food production will recoil from the massive scale of current production—abstaining from the takeover and biotic cleansing of large-scale landscapes and seascapes and limiting itself geographically so as to let Earth's diverse life overflow. Revamping the food system into an ecologically friendly subsystem of Earth has inexorable implications for the human population size: it points to a sustainable population in the ballpark of two billion (the global population roughly around a century ago). Revamping the food system also has inevitable repercussions for global trade, for an Earth-friendly food system will mean profoundly deemphasizing export-import food markets.

Humanity has an alternative choice to resigning itself to a population projection upward of ten billion, staying on the treadmill of ratcheting up food and commodities production and trade, struggling to manage an unpredictable climate and other adversities, while slugging through a pandemic-level heartache in the wake of pointless (nonhuman and human) death, suffering, devastation, exile, and extinction. Instead, we can move toward deindustrializing food, stabilizing and gradually reducing our global numbers, and learning how to live grounded in love and care of place and beings. Humanity will not advance by taking over the biosphere, but, on the contrary, will stagnate in the debased identity of the colonizer and decline in the conflict-ridden condition of jostling for "resources"—all the while clinging to its pathetic planetary dominance "only through the infinite management of its own collapse."[2]

There is still the road not taken. It will always be there, but the longer human beings cling to the delusion of their species' supremacy, the more irreversibly impoverished and downtrodden the biosphere to which humanity will finally humbly turn will be. Instead of later, we can choose now to live in loving fellowship with our earthly wild and domestic cohort, within vibrant ecologies, nestling human inhabitation inside the vast expanse of a living planet—allowing its exuberant dance of seasons, abundance, diversity, complexity, and evolution to resume.

PART ONE

The Destruction of Life and the Human Supremacy Complex

ONE

Unraveling Earth's Biodiversity

The living membrane we so recklessly destroy is existence itself.
JULIA WHITTY

There are so many stories narrated in scientific reports, naturalist and environmental writings, and the internet conveying the biodiversity holocaust. Tropics going up in smoke; grasslands turned over to monocultures; forests, coral reefs, savannahs, and steppes emptied of their animals; frogs, butterflies, bats, sea horses, freshwater fish, and honeybees blinking out; dwindling migrations; incalculable numbers of wild fish fished out of existence; plummeting populations of big carnivores and herbivores; elephants and rhinos gunned down by the thousands; coastal dead zones multiplying, and seas awash in plastic.

While each story demands attention in its own right, it is only by congregating them in our mind's eye that we can grasp the systemic scope of the crisis under way. Humanity is dismantling the very qualities that constitute the living world: variety of life-forms, complexity of life's interrelations, abundance of native beings and unique places on Earth, and diversity of nonhuman forms of awareness. These intertwined qualities form the cauldron of Earth's beauty and creativity. They are the ground of life's evolutionary power, fecundity, and endurance.

Biodiversity's facets of diverse life-forms, abundances of wild organisms, complexity of ecological relations, and variety of nonhuman lifeways may be described as the flame of life. Flame of life is a metaphor for life's richness at the levels of species, subspecies, populations, genes, behaviors, minds, and small- and large-scale ecologies—a richness that is self-perpetuating and builds more of itself over time. In the wake of the human onslaught, the flame of life is being extinguished. The richness of the living world is coming undone as the human juggernaut eclipses the stupendous diversity of our only cohort in the universe, turning the Earth into a biologically impoverished human colony and "stretching our loneliness to infinity."[1]

Biodiversity is disappearing because of the wholesale takeover of previously vast, connected, and free landscapes and seascapes, and the virtually unrestrained invasion into the planet's remaining wild nature. Wilderness, the matrix within which biodiversity thrives, is shrinking and becoming fragmented, resembling shards of natural areas in the midst of hostile developments such as industrial agriculture fields, grazing ranges, roads, highways, clear-cuts, mining projects, suburban sprawl, fences and other constructed barriers, and oil, coal, and gas ventures.

How big the human sea has become is captured by environmental analyst Vaclav Smil, who recently compared the biomass of wild vertebrate animals to the biomass of all humans and domestic animals. He found that "even the largest species of wild terrestrial vertebrates now have aggregate zoomass that is only a small fraction of the global anthropomass," and that "the zoomass of wild vertebrates is now vanishingly small compared to the biomass of domestic animals."[2] In brief, the combined weight of humans and domestic animals dwarfs that of the planet's remaining wild terrestrial animals. Smil's measure starkly captures the upshot of human expansionism—of population, economic, agricultural, and infrastructural growth. Humanity and its domestic animals have overtaken the biosphere, while wild creatures and places are dwindling. The destruction of life's diversity, complexity, and abundance profoundly downgrades the human understanding and experience of life's magnificence. As the living world is vandalized and its richness diminished, human beings become increasingly oblivious to the full spectrum of Earth's splendor.

Abundant Life

Imagine yourself one sunny morning in the late eighteenth century, standing on the shores of Wales and watching the undulations of a vast school of herring dodging a multitude of predators:

The arrival of the grand school [of herring] is easily announced, by the number of its greedy attendants, the gannet, the gull, the shark and the porpoise. When the main body is arrived, its breadth and depth is such as to alter the very appearance of the ocean. It is divided into distinct columns, of five or six miles in length, and three or four broad; while the water before them curls up, as if forced out of its bed. Sometimes they sink for the space of ten or fifteen minutes, then rise again to the surface; and, in bright weather, reflect a variety of splendid colors, like a field bespangled with purple, gold, and azure. . . . The whole water seems alive; and it is seen so black with them to a great distance, that the number seems inexhaustible. . . . Millions of enemies appear to thin their squadrons. The fin whale and the sperm whale swallow barrels at a yawn; the porpoise, the grampus, the shark, and the whole numerous tribe of dogfish, find them an easy prey, and desist from making war upon each other. . . . And the birds devour what quantities they please.[3]

This extraordinary display of marine wildlife was by no means exceptional, but typical of the biosphere's abundance of biological phenomena on land and seas.

Biodiversity is often misunderstood as referring to species numbers on Earth (or in any given ecosystem). This conception does a disservice to its meaning: numbers of species is a critical component of biodiversity, but biodiversity encompasses far more. The description above serves as an exhibit of its multilayered import. In the arrival of "the grand school," we discern a diverse cast of species and can infer the existence of many more. We also see huge numbers of animals and their relationships—an ecology in motion. The scene additionally points to the ways abundant life significantly shapes the environment: the erstwhile vast numbers of marine animals contributed to churning the seas vertically and horizontally, distributing nutrients and molding physical and chemical conditions. The feasting mass also tells us about emergent phenomena of interacting life-forms; marine biologist Callum Roberts writes that the appearance of the herring and their predators "ranked as one of the world's most remarkable wildlife phenomena."[4] We additionally glimpse another intrinsic quality of the ocean: its immense store of nutrients to support such seemingly "inexhaustible" numbers of herring[5]—even as the description depicts only one population of herring, while herring themselves are only one species among numerous other small fish.

The eighteenth-century author who encountered the grand school with its attendant millions of predators exclaimed that "the whole water seemed alive." Yet the scene intimates a bigger truth: the whole ocean was alive.

Even as this ensemble offers a view of biodiversity's many dimensions, so it serves as a window into biodiversity's crisis. The snapshot of life's richness conveyed in this eighteenth-century description suggests a baseline to begin to understand the profound declension that life is experiencing. A paramount aspect is the extinction of species that today is extremely high and heading the biosphere toward a mass extinction event. (I will elaborate on this shortly.) Yet equally significant dimensions of biodiversity destruction include plummeting numbers of wild organisms, the loss of wildlife and biological phenomena they give rise to, the diminishment of wild beings' ecological roles in food webs and nutrient circulation, and the eclipse of their contributions in cocreating complex biological, physical, and chemical environments.

The loss of such phenomena of abundance, as described above, also lifts the curtain on the colossal public ignorance surrounding biodiversity's unraveling. Ignorance about this momentous event has been conveyed through such expressions as "the declining ecological baseline," "the extinction of experience," and "ecological amnesia"—all ways of highlighting the collective obliviousness surrounding the eclipse of life's former richness.[6] Indeed, even as humanity is impoverishing the biosphere through species extinctions, extirpations of populations, unwinding ecologies, biological homogenization, and silencing of the polyphony of nonhuman lifeways, most people encountering such depauperate environments regard them as normal. Dimming knowledge and shriveling experiential horizons surrounding the wealth of planet Earth reveal how the human mind is afflicted by life's destruction.

The ongoing, cumulative forgetting of the biosphere's autochtonous nature is bringing humanity to the verge of losing the cosmic privilege of witnessing what can be neither fully comprehended nor, in the very long run, subdued: Earth's intrinsic being. Earth's being is a cosmos that self-creates itself through the resonances of its innumerable members, who, barring rare and large-scale catastrophes, keep swelling into a plenum of diversified kinds, abundant numbers, different ways of life, and exquisitely convoluted relationships—all unfurling as a slow-motion upsurge of biodiversity over geological time. The myriad, intertwining living elements scale up into the luminous mandala of the biosphere that we belong with. Life's multileveled diversity choreographs, even as it is shaped by, the inorganic dimensions of Earth. Earth is the most artful entity of the known universe, drawing wonderful compositions of life over unfathomable stretches of time.

Not that long ago, the salmon and other migrating fish of the Old and New Worlds fed the animals, the trees, and the soil, and their numbers still swelled to burst the rivers' seams. Before people turned the world into their *ecumene*—the human-dominated world—there were lions in Greece, cedars in Lebanon, whales in the Mediterranean, elephants in China, wolves in Japan, jaguars in North America, aurochs in Europe, bears in England, and temperate rain forests in Scotland and Ireland. There existed species of birds and fish that numbered in the billions. Herds of hundreds of thousands ungulates, including yaks, antelopes, gazelles, and wild asses, animated the Tibetan plateau. Cheetahs ranged from North Africa to India. One hundred thousand tigers roamed from the Caspian Sea through China and from Siberia to India, Sumatra, and Java. Whales abounded and ate krill in megatons, and the krill still proliferated to feed so many others. The carcasses and feces of millions of whales sustained a bizarre deep-sea life, one till recently unknown. Not that long ago immense numbers of sharks, swordfish, marlin, tuna, and other big fish traveled the ocean, and rainbows of living coral hallowed islands and coastal seas.

Rivers flowed free, nourishing some of the most life-abundant places on Earth within and around their waters. The world was filled with birds—seabirds, migrating birds, wading birds, flightless birds, huge and tiny birds, colorful and drab birds, vast flocks of birds, and raptors and scavengers with breathtaking wingspans. Massive herds trailed moving ecologies, plowing and fertilizing grasslands that overflowed in plenty, only to feed in turn the herds and their numerous, ever-in-motion attendants. Once the living world spoke to an Oglala Sioux named Black Elk, and he recorded the following: "I saw that the sacred hoop of my people was one of many hoops that made one circle, wide as daylight and as starlight, and in the center grew one mighty flowering tree to shelter all the children of one mother and one father. And I saw that it was holy."[7] The biosphere gutted of diverse, abundant life is not normal. For the biosphere, normal is abounding in endless, wonderful forms of life.[8]

Scientists do not often describe biodiversity in epic terms, but they do describe it comprehensively, as life's variety at the levels of species, genes, and ecologies. Roughly two million species have been discovered, with many more still undiscovered. The total figure remains unknown with estimates spanning between five and thirty million.[9] "We live on a little known planet," life scientist E. O. Wilson likes to say.[10] New species of worms, insects, fungi, plants, mammals, birds, reptiles, and fish are discovered all the time. For example, 10 percent of all known mammal species have been discovered since 1993; 26 percent of all known amphibians have been discovered roughly in the same period.[11]

Within species diversity lies even greater variety: the plush, gray zone of subspecies, varieties, and distinct populations that comprise a species. While "species" does refer to actual biological entities, the designation is, at the same time, an abstraction. In the real world, a species consists of a varied and dynamic tapestry of living beings. To illustrate with a couple examples, in pre-Columbian North America as many as half a million wolves enjoyed continental range.[12] Wolves roamed North America in distinct populations (called "metapopulations") and were composed of at least three subspecies. Africa and Asia were inhabited by millions of rhinos into historic times. The total numbers of the five species of rhinos today are estimated around thirty thousand, and three species—the black rhino, the Sumatran rhino, and the Javan rhino—are critically endangered.[13] Similar profiles were extant in many other species that existed in large numbers, wide ranges, different subspecies, and distinct populations prior to being decimated. Indeed, distinct populations of a species define a vital facet of biodiversity: metapopulation diversity at the global scale is, or more precisely was, enormous.[14]

Historic ranges, distinct metapopulations, total population numbers, subspecies, and varieties are all constitutive of the elliptical, monolithic notion of species. A more nuanced understanding of how species manifest opens a vista to the extraordinary phenomenology of life—to the experiential dimension of its bountiful manifestations. At a biological level, the above dimensions point to something equally vital though far more hidden: genetic diversity. According to biologists Rodolfo Dirzo and Peter Raven, a huge (and only partly explored) degree of rich genetic variation exists *within* populations as well as *between* them. "Many species," they add, "are composed of populations that are more or less genetically distinct from one another."[15] Genetic diversity is an indispensable ingredient of life, for it underpins the resilience of life-forms in the face of environmental flux. Variation at the genetic level enables nature to mold new life compositions, so that species are not only retained but also modified and diversified over geological time.

Biodiversity includes rich ecological variation as well. Diversity of places, composed of life assemblies that are morphologically and physiologically distinct, is described as Earth's *biodisperity*.[16] Continents and islands (both being types of landmasses separated hundreds of millions of years ago) became home to native, unique life lineages. Ecological diversity is also captured through the term "biome," which refers to large-scale ecologies comprised of interdependent plants, animals, fungi, and other life-forms within specific regions, climates, and altitudes. Examples include wet and dry tropical forests, temperate and boreal forests, deserts and alpine biomes, river watersheds and estuaries, grasslands and shrublands, tundra, continental shelves, and the high seas. Distinct ecologies are—or into historic times have been—homes for unique communities of beings.

Humanity is driving immense losses at all levels—species, genes (i.e., subspecies, varieties, overall numbers, and metapopulations), and ecologies. Life's diversity is in free fall on multiple fronts, through multiple causes, and in multiple places. The onslaught under way foreshadows the imminent reckoning of biodiversity destruction: humanity annihilating a living cornucopia that is self-replenishing and self-creative, and leaving in its wake a diminished, human-colonized planet. The systemic character of this onslaught impels us to recognize that if humanity chooses to keep life's flame ablaze, biodiversity collapse must be addressed at the scale it is occurring—piecemeal, bandage solutions will fall short.

The Systemic Assault on Life

The extinction crisis is a profound dimension of biological impoverishment. Because of life's evolutionary power (its ability to generate new life-forms) and its facility in spreading over the biosphere, biodiversity has steadily increased over time, beginning about 3.8 billion years ago and accelerating after the Cambrian explosion some 550 million years ago. Biodiversity's gradual swell has been virtually uninterrupted because the emergence of new species tends to be higher than the background (or natural) extinction of species.[17] This typical situation is annulled during episodes of mass extinction when catastrophic events, originating within or outside the planet, obliterate upward of 75 percent of Earth's species—no matter how well adapted or abundant those species may be.[18]

This describes the cataclysmic impact humanity is inflicting. The rate of extinction today is estimated to be one thousand times greater than the rate of background or natural extinction (the rate of extinction absent the human factor).[19] What's more, with life's destruction intensifying, the rate of extinction is expected to quicken in this century. Without concerted conservation efforts—and ultimately, without a profound shift in historical course—Wilson predicts that half the world's plants and animals will be gone by century's end. Others estimate that losses could reach two-thirds of Earth's species.[20] The biosphere is in the throes of a mass extinction episode called the Sixth Extinction because the geological record reveals the occurrence of five in the last 540 or so million years.[21] "Bracketed between best- and worst-case scenarios," writes naturalist Julia Whitty, "somewhere between 2.7 and 270 species are erased from existence every day. Including today."[22] Anthropogenic mass extinction is occurring rapidly—in geological time almost in an instant.

A circulating platitude that human-driven extinction is natural because extinction, as such, is natural is rehearsed in ignorance of the divergence between mass extinction and background extinction. Mass extinctions are extremely rare, global in scope, affecting the majority of life-forms, and demolishing biodiversity as a whole. None of these features apply to background extinction. Another extinction-is-natural cliché avers that extinction is so commonplace that the majority of species that have ever existed have gone extinct. This platitude regularly fails to add that a significant way species become "extinct" is by means of transmutation into novel ones. Mass extinction spasms, on the other hand, obliterate entire lineages of life-forms, terminating all possible evolutionary branchings. Charles Darwin called this type of extinction—when "no modified descendants" are left behind—utter extinction.[23]

"Extinction" is, misleadingly, the only word available for the death of species—irrespective of cause, magnitude, mode, or circumstance. The word's singularity obscures the fact that extinction does not refer to a uniform class of events. Mass and background extinctions, in particular, are entirely disparate phenomena. Background extinctions do not herald the wholesale decline of life, but are surpassed in magnitude by the birth of new life-forms. While in the case of background extinction the creative force of life prevails, with mass extinction obliteration rules. Mass extinction results in such a setback that it takes millions of years for the biosphere to generate a new chapter of biodiversity— a timeline that is meaningless for all human generations to come. If humanity does not seize the window of opportunity *now* to avert the extinction crisis, our legacy to all people will be a vastly and irreversibly impoverished planet. "We are thus engaged," as biologists Norman Myers and Andrew Knoll put it, "in by far the largest 'decision' ever taken by one human community on the unconsulted behalf of future societies."[24]

The significance of extinction can never be overstated because it means not only the death of a species, but the end of its evolutionary destiny as well—of the life-forms it would or might have originated as long as Earth orbits a life-sustaining star. Environmental ethicist Holmes Rolston rightly describes anthropogenic extinction as "a kind of superkilling." "It kills forms (species), beyond individuals," he continues. "It kills 'essences' beyond 'existences.'"[25] The extinction crisis today is not about species blinking out here and there but about a spasm of extirpations. Thus, all "extinction-is-natural" sophistry must be deposed. Anthropogenic extinction is anything but natural, for countless robust species are being extirpated in a blitzkrieg that happens rarely—as a matter of geological record roughly once in a hundred million years.

In the past handful of centuries, hundreds of species *that we know of* have been driven to extinction.[26] In the majority of cases, species that we have yet to discover (especially in tropical regions, freshwater ecologies, and islands) have been and continue to be annihilated. These Wilson calls anonymous extinctions, "not open wounds for all to see and rush to stanch but unfelt internal events, leakages from vital tissue out of sight."[27] "Many species," adds conservation biologist Stuart Pimm and colleagues, "will have gone or be going extinct before description."[28]

Plants, animals, and their communities are silently disappearing in places where endemism is high. Endemism is a critical dimension of biodiversity for it captures the unique feature of life-forms cloistered in certain places, found nowhere else, and often having a small range. Places with high degrees of endemism include the tropics, big and small islands, rivers, lakes, seamounts, and other special topographies, like the Everglades of North America and the Cape Floristic Region of South Africa to mention two. Endemic species are exceptionally vulnerable, because if their habitats are destroyed those species are annihilated. Most endemic species remain undiscovered. In the case of rain forests, for example, "some 19 of each 20 species [are] unknown . . . so the effects of burning rainforests result in a catastrophe beyond imagination."[29] (Meaning: tropical deforestation equals anonymous extinctions.) Many cases have been documented of recently discovered species driven to extinction just a few years later, indicating that countless undiscovered others have been extirpated before we knew them.[30] For "every species listed as endangered or extinct," according to biologist Bruce Wilcox, "a hundred more will probably disappear unrecorded."[31]

Life is in trouble everywhere not only where endemism is rife. Different kinds of losses are occurring in different places. High-endemism places are plagued by *species* losses. Temperate, boreal, and polar regions, as well as the ocean, are experiencing extinctions of *populations* faster than species extinctions. Indeed, a gigantic wave of biological annihilation is occurring at the level of populations, with millions of populations of wild beings wiped out worldwide every year.[32] And population extinctions are, of course, the prelude of species extinction.[33] Ecologies everywhere are under assault—especially wherever crops can be grown, domestic animals grazed, wild fish vacuumed, fossil fuels mined, human settlements located, and lucrative development unleashed.

Species become endangered by direct killing, or when their habitats are taken over, most often for animal and crop agriculture.[34] As land is increasingly appropriated, the original range of wild species shrinks, one population after another is obliterated, subspecies are extinguished, and typically all the above occur. Consider the case of the tiger. Until the turn of the twentieth century, tigers thrived in a variety of biomes and were abundant across Central and Southeast Asia. Massive reduction of their homelands (especially for agriculture) accelerated through the twentieth century. Today, there no longer exist large populations of tigers living in any one place. Alongside massive destruction of their habitats, tigers have also been killed relentlessly. Of eight subspecies, four have been terminated in recent decades: the Caspian tiger, the Bali tiger, the Javan tiger, and the Chinese tiger. In the span of one century, tiger numbers shrank by 97 percent. Adding insult to extreme injury, no wild tiger is safe today. As is the case for many other wild animals, poaching has recently increased, so that "tigers are now in critical condition everywhere, getting closer to extinction in the wild day by day."[35]

A virtually identical story can be told about cheetahs, lions, snow leopards, and jaguars, as well as other big animals, like orangutans, gorillas, sloth bears, pandas, wolves, whales, walruses, manatees, hippos, rhinos, giraffes, and elephants. Wild animals are disappearing—a calamity so conspicuous that scientists have named it "defaunation."[36] A recent study revealed that between 1970 and 2012 the world lost almost 60 percent of its wild animals. Animals of the world's lakes, rivers, and freshwater systems have declined by 81 percent.[37] These ominous findings are hardly surprising given the trends: drops between

30 and 90 percent have been reported for a wide range of species in recent decades, be it seabirds, parrots, freshwater turtles, sea turtles, sea horses, lions, primates, sun bears, pangolins, Tasmanian devils, bluefin tuna, gibbons, frogs, bats, or sharks. The list goes on. Thus, as appalling as extinction is in its finality, species can be devastated without being outright annihilated and places can be impoverished without being wiped out. The downfall of wild animals by means of massive killings and takeover of their habitats is unwinding ecologies, divesting the world of wildness and vitality, and wiping out lifeways, forms of awareness, and behaviors that mold topographies.[38] The energy of wild animals imbues the biosphere with dynamism. The systemic onslaught of defaunation heralds the de-animation of places and the silencing of terrestrial and marine expanses.

Drivers of Destruction: The Direct Causes

Two levels of causation are generally given to explain how the world has become hostile to biodiversity—direct and ultimate. The direct causes have been documented extensively: habitat destruction and fragmentation, killing, pollution, nonnative species, climate change, and the synergies between them. The ultimate causes—fueling the direct ones—are typically identified as human population size and growth, overconsumption, and technological power. Beyond the direct and ultimate drivers of life's destruction, I will argue that there exists a deeper causal layer: a human-supremacist worldview that sustains the conventionality of the direct hits and gives permission to the ultimate drivers to continue expanding. This work is dedicated to exposing and critiquing the worldview of human supremacy, and celebrating the restored Earth community that awaits its abolition.

Habitat destruction and fragmentation is the leading direct cause of biodiversity's decline, accounting for 70 percent of the plants and animals facing extinction.[39] In down-to-earth terms, "habitat destruction and fragmentation" refers to the human appropriation of geographical space across the globe—a takeover most zealously pursued when it comes to lush, productive, and readily accessible places.

When tropical forests are burned down or felled—for cattle ranching, soybean or sugarcane monocultures, oil palm plantations, subsistence farming, oil drilling, timber extraction, or mining—in one fell swoop species are wiped out or committed to extinction. Brazil, for example, has deforested about 20 percent of its Amazonia.[40] The Atlantic rain forest once covered an area four times the size of Great Britain; it's been reduced to 7 percent of its original area (mostly in fragments), for sugarcane cultivation, coffee plantations, and cattle ranches.[41] Many of its remaining species are endangered, while one-third of its tree species may go extinct due to the loss of seed dispersers.[42] Today, Southeast Asia has the highest rate of deforestation of all tropical regions (in good part due to oil palm plantations), and could lose 75 percent of its forests and half its biodiversity in this century if trends continue.[43] Haiti was once over 60 percent forested, but today 1 percent of woodlands remain, and its coastal seas are overfished.[44]

North America's largest biome, the prairie, is 98 percent gone.[45] The agricultural empire erected on this demolition exterminated the bison migrations (and almost the species itself), put numerous prairie life-forms on the endangered list, and of course annihilated the grasslands. In the span of the last two decades, continental-scale herbicide applications have wiped out almost one billion monarch butterflies by destroying their food.[46] ("The scale of loss is fantastic," states ecologist Chip Taylor, who heads the organization Monarch Watch.[47]) In Australia, habitat conversion to pasture and agricultural fields after European settlement contributed to the extinction of twenty-seven native mammal species.[48] The life-rich mangrove forests of tropical and subtropical coasts are being destroyed for aquaculture and coastal development, as well as for timber and fuel.[49] (Along the coasts of Central America, 40 percent of mangrove species are threatened with extinction.[50]) Industrial trawlers bulldoze the habitats of continental shelves and have taken to trawling the ocean's seamounts, devastating their living communities.

Habitat fragmentation is a form of habitat destruction, and it facilitates additional impacts.[51] Natural areas are fragmented by roads, highways, fences, power line and pipeline infrastructure, agriculture and pasture, settlements, and other development. In fragmented areas of boreal, temperate, and tropical regions, biodiversity is diminished by numerous effects: generalist species, including nonnative species and mesopredators, can breach such areas and outcompete (or devour) native species that require unbroken cover; pathogens find easier access, as do people with chainsaws, ploughs, and guns; and microclimates essential for many species are disrupted.[52] It is road construction penetrating into the Congo that has given access to poachers of forest elephants and other animals.[53] Deforesting and fragmenting Mexican woodlands where monarch butterflies winter allows freezing night air to penetrate their refuge, adding more threats to their survival.[54]

Fragments of wild nature lose population numbers and species over time. "Islands" of habitat are too small to support big animals, for example. In habitat fragments, insulated populations are also weakened from lack of genetic exchange with other populations. Numbers of plants and small animals decline when they cannot disperse across patches, becoming vulnerable to extinction. And as a species becomes whittled down to a few isolated populations, it can easily wink out from wildfires or disease.

Extensive knowledge about habitat destruction and fragmentation has led to the recognition that large size of natural areas, and connectivity between them, are critical for protecting biodiversity. The fact that destroying and fragmenting habitat is the leading cause of life's decline means, quite straightforwardly, that protected areas are indispensable for safeguarding wild nature and its inhabitants.[55] Were it not for protected areas, the extinction rate of mammals, birds, and amphibians would have been 20 percent higher.[56] Protected areas thus "deliver substantial outcomes for preventing extinctions."[57] But they are not foolproof—especially when protected areas are only "protected" on paper, as is the case for many (if not most) of Southeast Asia's nature reserves where the carnage of bushmeat poaching is emptying the region of all wild animals bigger than a tree shrew.[58]

Indeed, direct killing by guns, snares, poisons, hooks, and nets is exterminating staggering numbers of wild animals around the world. Recent research shows that direct killing is the primary threat to vertebrate species, especially the largest marine and terrestrial mammals.[59] Scientists have coined dismal neologisms to describe the killing fields: rifle extinction; defaunation; and empty forest, empty landscape, and empty coral reef syndrome. Animals are being killed (or captured in the case of the pet trade or entertainment industry), legally and illegally, for various motives: for food, for converting their habitats into agriculture, to protect livestock, for traditional medicine, for profit (turning body parts into commodities), or from fear. Industrial fishing is both a legal and illegal form of mass extermination that has devastated fish and other marine life. No one knows the exact amount of bycatch—dead and dying creatures thrown overboard—but it could be one-third of the landed catch and up to 80 or 90 percent in the most destructive cases.[60]

In developing countries around the globe, tens of millions of wild animals are slayed each year for subsistence purposes and market profits.[61] The bushmeat crisis involves a spectrum of animals from elephants and rhinos to primates, pangolins, and bats.[62] In Ghana, for instance, about 140,000 fruit bats are slaughtered every year; plant communities are dependent on the fruit bats to disperse their seeds, and so ecological tapestries come unstitched.[63] As hunting has gone from a relatively small-scale subsistence activity to a multibillion dollar international (often criminal-backed) trade, bushmeat is devastating wild animal populations and constitutes the most significant threat to the future of wildlife (especially mammals) in Africa, Asia, and South America.[64] Currently 301 mammal species are imperiled from hunting, three of which may already be extinct.[65] Population trends of hunted species have been worsening in the last few decades; only 2 percent of them have stable or increasing populations.[66] In a recent global overview of this critical problem, ecologist William Ripple and his coauthors conclude that "growing human populations, increasing middle-class wealth, access to hunting technologies in developing nations, and the modern ease of transporting goods around the planet are facilitating a global demand for wild animals for food and other products that simply cannot be met by current global wildlife populations."[67] Wild meat consumption is the primary but not sole cause; killing and trapping animals for medicinal products, ornamental purposes, and pets play a significant role as well. For example, the pet trade is menacing Indonesian and other Southeast Asian birds.[68]

Nonnative species (also called invasive, exotic, or alien) wield another direct hit on biodiversity. These are species that people transport, purposely or accidentally, into new environments. Nonnatives sometimes have no impact, and may become established inhabitants who fit in or even benefit their new homes (examples arguably include honeybees in North America and dingoes in Australia). But when nonnatives become literally "invasive," their effects are devastating. Domestic cats, for example, kill between 1.4 and 3.7 billion birds every year and even greater numbers of small mammals.[69] The brown tree snake accidentally introduced to the Pacific island of Guam devoured many of its native birds and bats out of existence. The Nile perch purposely introduced into Lake Victoria annihilated half its endemic fish. The red fire ant has spread from South America to North America and other parts of the world, displacing native insects and reducing wildlife populations.[70]

The introduction of rats, goats, and pigs has devastated countless native animals and plants especially on islands. Such nonnatives extinguished twenty-six plant species, eleven bird species, and the extensive forests of the Mexican Guadalupe Island.[71] Avian flu, transmitted by an invasive mosquito, has driven sixty Hawaiian bird species to extinction or decline, and threatens more.[72] The chytrid fungus that is wiping out frogs worldwide has been spread by people. The proportions of species threatened because of introduced animals and plants are, in some cases, enormous. For example, most threatened species in Hawaii are endangered from nonnatives. In the United States, between 25 and 40 percent of threats are due to invasive species.[73] Generally, nonnative species "have been solely responsible for 20 percent of extinctions since 1600 and partly responsible for half of them."[74]

Pollution afflicts the entire biosphere, sickening and killing beings as well as obliterating ecologies. There are four hundred dead zones worldwide: estuaries that end up oxygen starved as a consequence of fertilizer and pesticide runoff.[75] Noise pollution and chemical contamination contributed to the extinction of the freshwater Yangtze dolphin. (The remaining five species of freshwater dolphins are all endangered by multiple threats.[76]) Most of China's rivers and lakes are polluted with agricultural chemicals and industrial waste—the collateral damage of rising economic prosperity coupled with a huge human population. Many rivers around the world are threatened by the same fate as China's rivers, where "80 percent of the 50,000 kilometers of major channels can no longer support fish of any kind."[77] Pollution is responsible for 20 percent of all species threats in China.[78]

On the other side of the globe, according to a 2014 scientific publication, "one in four of the St. Lawrence whales are dying of cancer, mostly intestinal cancer." The report adds that "epidemics of liver cancer have been found in 16 species of fish in 25 different polluted freshwater and saltwater locations. The same cancers were found in bottom-feeding fishes in industrialized and urbanized areas."[79] Toxic pollutants have invaded everywhere, including the Arctic where hazardous chemicals are compromising the reproductive systems of polar bears. Dolphins and killer whales have high levels of PCBs (polychlorinated biphenyls) in their bodies. The contamination is compromising the animals' reproductive systems, so that European killer whale populations, for example, are facing extinction.[80] Across North America, as elsewhere, light pollution lures and causes the death of millions of migratory birds through collisions with buildings and other structures.[81]

The seas are filling with plastic—from plastic bags, tiles, sheets, and lost fishnets forming the Great Pacific Garbage Patch (and other small "continents" of trash) to invisible-to-the-eye microplastics.[82] Plastic is killing marine mammals, sea turtles, fish, and seabirds. Among seabirds, the surface-feeding petrels, shearwaters, and albatrosses, as well as their fledglings, are most afflicted.[83] Chicks perish in their nests, their decomposed bodies a grotesque exhibit of colorful plastic trinkets that had been fed to them, haloed by a ring of downy feathers.[84] Plastic bags strangle the intestinal tracts of whales and sea turtles. A sperm whale beached in 2012 was found to have in his stomach thirty square meters of tarpaulin, over four meters of long hose, a nine-meter plastic rope, and two flowerpots.[85] All told, a million seabirds, one hundred thousand marine mammals, and countless fish die in the ocean each year from ingesting plastic or getting ensnared in it.[86] Microplastics are entering the ocean food web with unknown repercussions for non-human and human life. A meta-analysis of one hundred peer-reviewed papers on plastic contamination concluded that all marine groups are at high risk of consuming microplastics.[87]

Carbon dioxide was classified as a global pollutant when scientists realized that its accumulation in the atmosphere is changing the climate and causing ocean acidification. Both atmospheric carbon dioxide and climate change occur naturally, but the speed at which greenhouse gases are mounting and climate is shifting leave nature's typically far slower pace in the dust. If carbon-loading the atmosphere continues apace, the biosphere is heading toward an average temperature differential as sizeable as that between glacial and interglacial periods—but at a tempo ten times faster. The natural world has no time to adjust, and terra incognita looms on the horizon of all complex life. Studies indicate that one in six species could face extinction from climate change alone.[88] Alpine ecologies, tropical forests, coastal mangroves, the tundra, and coral reefs, among other places, are at risk. Across North America's Rockies stretch millions of acres of dead trees, killed by the mountain pine beetle that warmer temperatures favor. More generally, big trees are dying: the density of large-diameter trees—the largest living beings and among the oldest—has fallen by 25 percent.[89] Many amphibian extinctions are tied to a spreading disease that global warming encourages, illustrating the more broadly applicable point that "climate-driven epidemics are an immediate threat to biodiversity."[90]

Earth's biggest mass extinction event—in the Permian-Triassic when 90 percent of marine species and 70 percent of terrestrial species died off—may have occurred from a global warming episode in which sea and permafrost methane deposits were released, triggering runaway heating. It is thus abundantly clear that anthropogenic climate change has the potential, all by itself, "to run Genesis backward," as environmental author and climate activist Bill McKibben puts it.[91]

The seas are already 30 percent more acid than they would have been without greenhouse emissions—more acidic than they have been in eight hundred thousand years.[92] Acidification threatens the ocean's entire food chain. Certain algae that dwell in tiny shell structures (called coccolithophores) are consumed by copepods (among other organisms), who in turn feed small fish and krill, who are vital for tuna, dolphins, whales, seabirds, and other marine life. The threat of coccolithophores declining, from their shells corroding, can thus perturb the entire web of life. Ocean acidification endangers other calcifying creatures (organisms that build a shell or external skeleton from calcium carbonate), many of which create habitats with their bodies, like coral reefs, oysters, barnacles, and certain seaweeds. Coral reefs are critically endangered from both global warming and ocean acidification. The symbiotic partnership forming the living coral comes undone at a certain level of sea warming, while the calcification process through which the coral body creates itself becomes derailed at a certain level of acidification. Being profoundly biodiverse ecologies, deteriorating and dying reefs translate into losses of countless beings who depend on them for food, shelter, and reproduction. If humanity continues the business of adding carbon to the atmosphere, "coral reefs will cease to exist as physical structures by 2100," or earlier.[93]

Species have "climate envelopes," meaning a range of temperature and hydrological conditions they are adapted to. When these conditions change, there are three possibilities: species adapt, move, or perish.[94] When climatic patterns shift substantially, animals and plants largely respond by moving. But in our time, life's movement is severely hampered. Scientists Terry Root and Stephen Schneider note that the synergy between habitat fragmentation and climate change poses a grave challenge. "As the climate warms," they write, "individual species of plants and animals will be forced to adjust if they can. . . . During the Ice Age transition many species survived by moving to appropriate habitats. Today such dispersal is more difficult because they need to travel across freeways, agricultural areas, industrial parks, and cities."[95] Many life-forms will perish from anthropogenic climate change because they are barred from moving, have nowhere to move to, move too slowly, and/or have very sensitive "thermal limits." Herpetofauna (reptiles and amphibians) have both sensitive thermal limits and move slowly. A rapidly shifting climate will be hard on these creatures. A 2010 study estimated that 20 percent of all lizard species could be wiped out by climate change by 2080; for those who survive their numbers will dwindle.[96]

I have endeavored to canvass, with examples, the direct hits on biodiversity: habitat destruction and fragmentation, killing, nonnative species, pollution, climate change, and ocean acidification. Each of these causes is formidable in itself, yet the most lethal blow to life comes from their synergies. Recent research into the causes of species' threats found that more than 80 percent of those studied are afflicted by more than one major threat.[97] It is because of synergies between multiple hits that the biosphere is in the grip of a mass extinction episode. (Synergies between multiple causes are also thought to have been decisive in the previous five.[98]) Research into the condition of coral reefs worldwide makes the point starkly: Reefs are critically endangered in the Caribbean, where overfishing, coastal development, pollution, disease, ship and boat anchors, and tourism—along with warming and acidifying seas—have all but destroyed them. But in the Pacific atolls, where "only" the impacts of global warming are being experienced, coral reefs are faring better.[99]

Ditto for amphibians. Habitat loss to agriculture and other developments, pesticide pollution, ozone-layer depletion, climate change (affecting hydrological conditions), competition with nonnatives, and an invasive organism (the chytrid fungus previously mentioned) are colluding to make them, along with freshwater species, among the most endangered creatures on Earth. Seventy species of South American frogs have gone extinct in the last twenty years.[100] Overall, more than 40 percent of amphibian species are endangered.[101] That amphibians emerged roughly three hundred million years ago, and survived the asteroid impact that occurred sixty-five million years ago, brings home the magnitude of humanity's assault on their being.

Turning to other examples: All the kiwi species of New Zealand, a flightless group of birds, are endangered from a coalition of factors—from introduced species like cats and dogs, to habitat destruction and roadkill.[102] Southeast Asia is predicted to lose 40–50 percent of its species by 2100 from deforestation for agriculture and timber, bushmeat killing, wildlife trade, and increased numbers of human-driven wildfires.[103] (The region of Southeast Asia includes Thailand, Laos, Cambodia, Myanmar, Indonesia, and Malaysia.) The world's penguins are also in trouble from more than one cause: climate change is altering their habitats while industrial fishing is draining their food supply.[104] Many bat species are imperiled from introduced disease, invasive snakes, habitat destruction, bushmeat hunting, and attacks on their colonies.[105] Multiple threats are endangering an estimated one-third of land plants and 68 percent of flowering plants.[106] Innumerable Earth creatures face this lethal pattern of "a one-two punch."[107]

The destruction of freshwater life is also a consequence of multiple hits. According to researcher Zeb Hogan, "globally, a pattern has emerged: large aquatic animals are disappearing. The world's river dolphins and large freshwater fish face the biggest threats, including overfishing, dams, navigation projects, pollution, and habitat destruction."[108] Imperiled freshwater animals include the critically endangered gharial (a freshwater crocodile once abundant in the rivers of India and Nepal), the wild Chinese alligator, the black finless porpoise, the Yangtze giant softshell turtle, and the giant salamander, among others worldwide.[109] "Unless concrete steps are taken soon to better protect these vulnerable species," Hogan warns, "this is the beginning of a wave of extinctions that is likely to occur over the next 20 to 30 years."[110] And it is not only large freshwater animals who are imperiled. According to the Millennium Ecosystem Assessment, freshwater ecosystems have the highest proportion of endangered species. Thirty percent of all freshwater reptiles are threatened, while the proportion rises to 50 percent in the case of freshwater turtles, who are also plagued by national and international trade.[111] Twenty percent of Africa's lake species are endangered. The culprits are the usual legion: agriculture, deforestation, water extraction, damming, pollution, and introduced species.[112]

The synergy between habitat fragmentation and killing is particularly insidious, as the containment of animals in fragmented natural areas makes them vulnerable to poaching. Thus, though protected areas are indispensable for safeguarding the remaining populations of threatened animals, unless such areas are vigilantly guarded they become death traps—the very places that poachers know to find the animals. With wildlife populations rapidly declining outside protected areas, poachers are moving into reserves and parks.[113] Insulated islands of nature are unsafe zones for animals. Adverse synergies also manifest in surprising ways. For example, when I was in Madagascar over ten years ago, researchers studying its endangered lemurs lamented that the fossa—a native catlike carnivore itself classified as "vulnerable"—would move into forest fragments and pick out its resident lemurs, like so many "sitting ducks," sometimes in a single night.

In sum, life's diversity is threatened at the levels of species, genes, and ecologies. "Nature's big shows,"[114] like migrations and wildlife spectacles, are vanishing. The evolutionary potential of complex life, especially of large vertebrates, is being stymied or suspended.[115] Entire classes of beings are under fire: big wild animals, freshwater species, amphibians, seabirds, bats, pollinators, and flowering plants. Entire biomes are being diminished or eliminated, along with their exceptional diversity and complexity: tropical forests, grasslands, wetlands, estuaries, rivers and lakes, coral reefs, continental shelves. All told, complex qualities, distinct kinds, and vast numbers of wild beings and places are being eclipsed. The direct hits are multiple, and most threatened organisms face more than one of them. No place is safe from the heavy hand of depredation, from pole to pole and from the Mariana Trench to the Himalayas.[116]

Describing the onslaught on life as *systemic* is a way of organizing and clarifying the dizzying number of reports streaming in about biodiversity destruction over the last few decades. The point of the exercise is not to encourage despondency. Nor is to underrate the heroic efforts of conservationists, scientists, and ordinary folk working to protect life, and their significant victories worldwide.[117] The point of comprehending the magnitude of life's ruin is simply to recognize the corresponding magnitude of the shifts needed for Earth to remain a living planet.

The systemic character of life's undoing might be described as the anti-epic of our time. An epic is "an impressively great" endeavor recounted in "a long poetic composition, usually centered upon a hero, in which a series of great achievements or events is narrated."[118] The anti-epic of biodiversity's destruction is the unraveling of Earth's poetry. There is zero greatness in it—hence a big public silence shrouds it. For those who ponder it, it seems natural enough to suspect that an "antihero," wielding multiple weapons, is the culprit: Who could the villain be but *Homo sapiens*? A central argument in this work, however, is that the human species is not the villain. No original sin hounds us, no deep-seated, antilife defect is etched into our genome.

In agreement with author Paul Hawken, "we can see the world as doomed and fatally flawed or we can see every trend and statistic as a possibility of transformation." "We must change," he continues:

In this century we can commence the work of ecological and cultural restoration on a grand scale. We can begin to reduce carbon in the atmosphere; recharge aquifers; restore lands that have been taken by deserts; create habitat linkages for buffalo, panthers, and wolves; and begin to rebuild paper-thin soil. We can create a world where wilderness coexists with civilization. We can and will do this because it is the only way we can be fully human, and the only way Earth's grace will sustain us.[119]

Drivers of Destruction: The Ultimate Causes

The direct drivers of biodiversity destruction have been documented extensively. There's also convergence regarding the ultimate causes: too many people, consuming too much, and using destructive technologies. These underlying drivers are often summarized in the formula IPAT: environmental impact (I) as a factor of population (P) size, affluence (A), and technological (T) capacity.[120] While the gigantic blow to biodiversity appears adequately captured in the convergence of PAT, by itself this metric is explanatorily insufficient. It cannot explain *why*, especially in this time of planetary emergency, little is being done to rein in the ultimate drivers of life's undoing. IPAT remains silent about an unquantifiable, deeper cause of biodiversity's unraveling: a shared worldview that authorizes humanity to continue expanding its numbers, appetites, and technologies. The worldview of human supremacy, as I will elaborate throughout this work, champions and normalizes human expansionism. As long as it prevails, humanity will remain unable to muster the will to scale down and pull back the burgeoning human enterprise that is unraveling Earth's biological wealth.

In today's globalizing world, humanity's large and growing population, escalating consumerism (especially with a rapidly rising global middle class), and spreading (mega) technologies amount to a unified front that is reconfiguring the biosphere into a human-dominated, biologically sheared planet. One way to cinch the detrimental congruence of these three factors is to foreground their synergies. The domains in which human numbers, affluence, and technology synergize to deliver the biggest blow to life on Earth are food production and global trade. Indeed, industrial food and international trade easily qualify as the arch-offenders of Earth's integrity. Industrial food production on land and seas is, hands down, the most ecologically devastating activity. The international trade of food, along with the trade of huge quantities of other commodities, compounds the destructiveness of the industrial food system, erodes human loyalty to place, and encourages the overproduction of everything—including animal feed, food products, junk stuff, luxury items, and throwaway commodities. The planetwide damage of these two human systems involves the convergence of huge numbers of people, excessive patterns of consumption, and technological assemblies that are, no metaphor intended, weapons of mass destruction.

While enough food is grown and harvested to support the 7.5 billion people alive today—even as eight hundred million continue to be hungry, malnourished, and/or food insecure[121]—the cost to the natural world has been colossal. Food production occupies roughly 40 percent of Earth's ice-free land. (The conversion of land to almost exclusive human use could reach 70 percent within this century under business-as-usual growth scenarios.[122]) Today, an area the size of South America is devoted to crop cultivation and an even larger area, the size of Africa, is allotted to livestock grazing.[123] Pasture occupies around two-thirds of the world's total agricultural land.[124]

Domestic farm animals and cultivated crops comprise but a fraction of the world's species, yet they have devoured the lion's share of geographical territories and biomass. Thus are wild animals, plants, ecologies, and biological phenomena crowded out of existence by vast stretches of monocultures, by billions of farm animals, and by humanity's overall appropriation of the biosphere. Land occupation for animal and crop agriculture is the chief cause lurking behind what conservation biologists call "habitat loss" and "habitat fragmentation," which, along with direct killing of wild animals, continue to be the leading destroyers of biodiversity.[125]

Ninety-eight percent of the land where rice, wheat, and corn can be grown has been converted for that purpose.[126] Not surprisingly, around 80 percent of freshwater diverted for human use is for agriculture.[127] Agriculture, to rehearse a well-known statistic, contributes more greenhouse emissions than the transportation sector. Including the effects of deforestation for agriculture, at least 25 percent of anthropogenic greenhouse gases stem from food production, with emissions from the livestock sector accounting for 80 percent of that figure.[128] (These emissions are driving a climate-change episode that, barring a drastic energy transition or carbon sequestration solution, could edge the planet toward an average temperature increase in the ballpark of the Paleocene-Eocene Thermal Maximum.) Agriculture is also responsible for the world's four hundred dead zones, which have been doubling in number every decade since the 1960s, tracking the spread of Green Revolution technologies and livestock and human population growth.[129] Runoff from agricultural fields and confined animal operations also strangles freshwater lakes, rivers, wetlands, and streams worldwide.[130] Food production drives soil erosion and desertification. The industrial production model of monocultures requires applications of herbicides, insecticides, and other biocides. Indeed, many consumers and growers alike have been persuaded by corporate spokesmen and governmental agencies to regard poisoning the biosphere—in the name of producing human nourishment—as normal or necessary.

Many of Earth's small-scale to large-scale ecologies have been dismantled by food production. The temperate zone has been effectively given over to agriculture. Temperate grasslands are probably the hardest hit, with only 2 percent remaining in a natural state as noted earlier.[131] Since the 1980s industrial agriculture has also moved into the tropics, extinguishing untold numbers of species. Most tropical deforestation in the last three decades is directly beholden to the expansion of crop plantations and ranch operations.[132] Over half the world's species-rich wetlands have been drained over the last century largely for repurposing into agriculture.[133] Mangrove areas are rapidly declining with aquaculture operations being a major driver.[134] An estimated ten thousand to twenty thousand freshwater species are at risk of extinction, and river biodiversity is most threatened in regions of intensive agriculture and dense settlement.[135] The factor of agriculture accounts for 76 percent of primates threatened with extinction.[136] Coastal seas are critically endangered and continental shelves are endangered primarily because of overfishing.[137] Treasures of seamount life and habitats are being bulldozed by trawling vessels.[138] Indeed, the two least disturbed biomes on Earth—boreal forests and tundra—are, tellingly, two biomes where large-scale food production does not (yet) occur.[139]

Ninety-seven percent of the ocean is legally open to fishing. Marine life and ecologies have been pillaged—globally by industrial fishing and regionally by agricultural and other forms of pollution. The composite impact of human population size, a rising global middle class, international trade of seafood, and technological gigantism and state-of-the-art fishing technologies have teamed up in the plunder of wild fish: the majority of commercial "fisheries" are depleted or unsustainably fished, most big fish are gone (fish such as sharks, swordfish, tuna, and marlin), and large-scale aquaculture, livestock operations, and the pet industry contribute to the depletion of the ocean's small fish for feed.[140]

On land, declines of big animals are also directly linked to food production. The former ranges of big herbivores have been diminished through competition with livestock and by conversion of their habitats into cultivated lands.[141] Former abundances of the world's large herbivores are no more, and associated phenomena of their mass migrations are also disappearing.[142] Large carnivores are killed because they threaten farm animals, and they are also harmed because their prey species are shrinking.[143]

Briefly put, ecological demolition is being sponsored by breakfast, lunch, and dinner plates around the world. Economist Jeffrey Sachs notes that "agriculture is the main driver of most ecological problems on the planet," while ecologist Jonathan Foley observes that "pasture has become the dominant ecosystem [*sic*] on the planet."[144] The massive conversion of land for large-scale animal and crop agriculture and the wholesale onslaught on marine life by industrial fishing constitute the chief agents of habitat destruction; species extinctions; wild animal population declines; global, regional, and local pollution; rapid climate change; and the overall dismantling of ecologies. The extractive activities of logging and mining, and the incessant construction of infrastructures like roads, pipelines, cell towers, and dams, are also destructive—and often associated with food production and the facilitating of global trade.

Human societies worldwide are emulating two industrial food trends that are devastating life: the mass consumption and trade of meat (and other animal products) and feedstock; and the overconsumption of processed foods, packaged foods, and beverages, which are energy and materials intensive, trucked, flown, and shipped around the world, and overrunning terrestrial and marine environments with trash.

With the simultaneous growth of the global population and consumer affluence, the consumption of meat, fish, and other animal products has been accelerating worldwide.[145] Since 1950, each person on average eats twice the meat and four times the seafood.[146] This decades-on-the-rise trend continues. In our time, notes journalist Mark Bittman, animal-derived foods demand "so many resources that it's a challenge to enumerate them all."[147] Author Tony Weis has gone to great lengths to lay out the massive impact of the industrial-livestock complex, which continues to globalize without end in sight. He summarizes the price of this industry as involving "tremendous amounts of land, water, fertilizer, chemicals, fossil energy, toxic runoff, nutrient loading, and greenhouse gas emissions."[148] This formidable list—each item a litany in its own right—does not include the industry's ethical price: the debasement of human integrity in the maltreatment of once life-filled terrestrial and marine ecologies, as well as in the abuse of farm animals who are put through an industrial production system and mass slaughtered without being cared for or ever thanked.

The production of animal-derived foods, with little to no sense of limitation, is propelling a biological holocaust on Earth for a number of reasons: these foods require the conversion of huge swaths of wild nature into cropland for feed;[149] they demand the repurposing of vast territories for grazing; they foster the killing of carnivores to protect livestock;[150] they displace wild herbivores whose homes are pilfered for conversion to fields and pasture; they are responsible for the potent greenhouse gases methane and nitrous oxide that come from farm animals; and they undergird the devastation of marine life by industrial fishing, alongside the regional despoliation of coastal seas by industrial aquaculture. If we add to this mix the hunting of wild animals for meat—the ruination known as the bushmeat crisis—we can easily recognize that "increasing levels of human carnivory are the crux of the problem."[151]

The growth of the global farm animal population—both grazing and confined—has resulted in escalating air, land, and waterway pollution, as well as in the desertification of grasslands and shrublands (as, for example, in Patagonia, Mongolia, and sub-Saharan Africa). South American rain forests and dry tropical forests are destroyed to raise cattle and to grow soybeans for animal feed. The greatest portion of the corn and soybeans grown in the United States is also for confined animal feedstock. Overall, in developed nations up to two-thirds of total grain production is devoted to feed. A quarter of the world's fish catch is fed to livestock and to carnivorous fish in aquaculture operations. In the developing world, land needed for grazing and browsing wild animals is given over to livestock whose numbers *tripled* between 1980 and 2002, tracking both enormous population growth and a globalizing Western diet.[152]

The blown-out-of-all-reasonable-scale livestock sector—socially engineered by special interests targeting a growing consumer population with cheap animal products—is devastating biodiversity, cruel to the farm animals, and unhealthy for people. It is "essentially indefensible," in McKibben's words, "ethically, ecologically, and otherwise."[153] This essentially indefensible sector is projected to double by 2050 (over 2010 levels)—and under global business as usual there's no reason to believe that it will stop growing then.

The massive impact of industrial food production, and its livestock and industrial fishing sectors in particular, is one adverse synergy of huge numbers of people with growing purchasing power. The massive impact of global trade—in which trading food features big-time—is another domain in which the variables of "population" and "affluence" are inseparably meshed.

The connection between export commodities (especially cash crops) and ecological degradation has long been known on a case-by-case basis. But a study published in *Nature* in 2012 provided a comprehensive picture of international trade as a driver of life's destruction.[154] The study linked twenty-five thousand endangered species to fifteen thousand commodities produced in 187 countries: in other words, it connected the dots between threatened species lists and global trade databases. This was the first research to quantify the role of trade in biodiversity losses, revealing that a significant portion of species threats "are driven by economic activity and consumer demand across the world."[155] Local and regional extirpations of populations, species, and ecologies are thus beholden to a global economy that prioritizes export-import relations between billions of people.

The study disclosed that for the developed world—the United States, the European Union, and Japan, for example—a large portion of their detrimental impact on biodiversity is due to importing commodities from outside their borders. In a mirror image, developing countries degrade the lands and waters within their borders for the sake of producing exports. On average, 30 percent of species threats are due to the global trade of common food products like coffee, sugar, cocoa, tea, bananas, meat, fish, soybeans, and palm oil, as well as nonfood products such as rubber, textiles, lumber, and manufactured goods. For developing countries (for example, Madagascar, Papua New Guinea, Sri Lanka, and Honduras), the number is as high as 50–60 percent of species threats within their borders due to global trade. Consumers in developed countries, the *Nature* article reports, are "the largest beneficiaries" of these trade flows. Ripple's work further corroborates this fact: he and his colleagues found that a substantial threat to large animals in Latin America, Africa, and Southeast Asia is habitat conversion to create agricultural and other products demanded by the developed world.[156] Much biodiversity destruction thus stems from a social order built by political and economic arrangements that favor international trade. This order facilitates the virtually unrestricted flow of commodities directly at the expense of the natural world.

Understanding global trade's dark role in the demolition of life gives the lie to a circulating platitude that economic growth and development (as they are presently carried out) are "good for nature," because after countries develop they have the luxury to turn their attention to conservation and environmental causes. "More growth, not less, is the best hope for averting a sixth great extinction," opines, for example, the *Economist*.[157] In a similar vein environmental scientist Erle Ellis writes that "as populations, consumption, and technological power advance at an exponential pace, industrial systems appear to be evolving in new directions that tend to reverse many of the environmental impacts caused by agriculture and prior human systems."[158] Contrary to such claims, something entirely different occurs after countries develop economically and their middle- and upper-class populations swell. Such countries are able to spare a portion of their own places from development, because they acquire the wealth and power to flip their nature-destroying footprint to poorer and less powerful countries elsewhere.

As author Alan Weisman puts it, "rich countries fly high on the wings of distant lands."[159] At the same time, the citizens of rich countries have the luxury and profusion of distractions to disregard, in environmental philosopher Val Plumwood's words, "the many unrecognized, shadow places that provide our material and ecological support, most of which, in a global market, are likely to elude our knowledge and responsibility."[160] We are left to ponder the fate of the biosphere as billions of people join the consumer ranks—something already under way and clearly the aim of mainstream development efforts. Biodiversity destruction will then become an equal opportunity venture of a global consumer society: a pact of mutual nature destruction among equally developed nations, driven by an economic order oriented primarily toward export-import markets to serve billions with a profusion of food products and other commodities.

The authors of the *Nature* article were compelled to offer policy recommendations for alleviating pressures on species and ecologies due to trade. They propose "suppressing trade in at-risk commodities," while correctly observing, a sentence later, that "such a policy reform would be difficult to implement given the importance of international trade."[161] Indeed, the commodities responsible for wiping out and endangering wild nature are worldwide consumer staples—especially animal feed and food staples. The possibility of suppressing their trade is not only absent on the immediate horizon, but the volume of global trade is expected to balloon with the further growth of the human population—and of its global middle class in particular—facilitated by economic connectivity and the expansion of infrastructural networks of roads, highways, canals, seaports, airports, railways, warehouses, and global communications.

Burgeoning trade is completely reliant on infrastructures, and the trends in coming infrastructure development are far from benign. Global plans are at hand to invest trillions of dollars into new mega-infrastructure projects in the transportation, energy, agriculture, and water sectors. Meaning: more highways and roads, pipelines, electric power plants and electricity grids, mega-dams, communication systems, new and expanded seaports, and gigantic irrigation schemes for water transfers.[162] All such projects are ecologically harmful in their own right, and they will additionally amplify the nature-destroying activity of global trade among billions of people.

The *Economist* profiled one seaport, Long Beach, California, as "offering a prism on the global economy."[163] Its container traffic has more than doubled in the last twenty years, but even so its current capacity is considered inadequate because another doubling of traffic is expected within a decade. Future projections have led Long Beach, as well as the Los Angeles port, to plan spending $5 billion on infrastructure in order to accommodate larger container ships than they do today.

This prism on global trade considers just two ports. But worldwide, seaports, airports, and road systems are all pressed by global trade's growth imperative. For example, the Panama Canal is slated for expansion to accommodate vessels twice as large as presently allowed, while the opening of Arctic Sea routes, with the prospect of ice-free summers, will be amply exploited by the global shipping industry. Enlarging our prism further, we can mull over the ecological implications of Africa becoming "China's biggest trade partner."[164] We might also consider, borrowing the non-ironic words of the *Economist*, that trade between China and Brazil has become "increasingly vital."[165] Part of this "vital" trade consists in trading off rain forests for cattle ranches and soybean fields to produce beef and feedstock for export. Soybeans from the Brazil's rain forest end up in China's hog factory farms for the production of industrial pork, amounting to a trade chain that is hemorrhaging vitality on every front—devastating wild nature, farm animals, and human health.

For the Sake of Soybeans and Palm Oil

The impacts of overpopulation and mass consumption are inseparable, as evidenced in their synergies of industrial food production and global trade. Brief consideration of the commodity chains of soybeans and palm oil can serve to drive the point home. Soybeans and palm oil are among the biggest enemies of life on Earth, and their production and trade are beholden to humanity's huge population size, growing affluence, and the rising convergence between the two.

The largest soybean producers today are the United States, Brazil, and Argentina. The majority of the global soybean crop is turned into feed for livestock. The greatest importer is China: their soybean imports grew from $75 million in 1995 to $38 billion in 2013—an increase of several orders of magnitude in less than ten years—supporting China's "livestock boom," especially pork production.[166] (The pig population in China is twice the population of Americans.) Let's translate this demographic-economic picture into ecological terms. In the United States, the largest portion of the prairie biome (which might still be restored and rewilded, as I will argue later) has been obliterated in order to grow herbicide-resistant and herbicide-drenched soybeans for feed. (Soybeans are rotated with corn, another industrial crop turned largely into feed.) All but gone are the North American grasslands with their extravagant community of life, from grizzlies and pronghorn antelope to bison, prairie dogs, and thousands of plant species. Beyond the United States, species are daily driven to extinction in Amazonia for the sake of feedstock, while the pampas of Argentina are obliterated—all for soybeans needed to mass-produce meat and dairy.

The destructiveness of global food production and trade is not limited to rising meat, dairy, and seafood production. Another traded product for which rain forests are falling is palm oil.[167] In recent years, palm oil has been steadily added to more and more processed and packaged foods, including many (if not most) vegetarian and vegan foods. Indeed, palm oil is the new corn syrup, making a consumer debut as an added ingredient in veggie burgers, vegan cheeses and butters, ice creams, cookies, margarines, crackers, potato chips, peanut butters, and more. Palm oil is also increasingly added to soaps, candles, shampoos, and cosmetics, and turned into biodiesel that is shipped to Europe and elsewhere. Over 80 percent of the world's palm oil is grown in Indonesia and Malaysia, where, for the sake of that export crop, forests of stupendous biodiversity are being destroyed.[168] Ecologist Gerardo Ceballos and his coauthors rightly call this a "holocaust of nature."[169] The population of Indonesia's endangered orangutans and proboscis monkeys, for example, have declined steeply in recent years as a consequence of expanding oil palm plantations.[170] In Malaysia, 266 freshwater fish species have been driven to extinction.[171] With the global consumer class growing, the market for palm oil is expected to increase substantially. There is a real possibility of the lucrative venture and ecological scourge of industrial palm oil expanding into African and South American tropical forests.[172]

Because of their industrial-scale production and global-trading scope, soybeans and palm oil are important examples; it behooves us to re-member, however, that they are only two examples. (The explosive in-dustrial growth and trade of apparel is another important one.[173]) We can extrapolate beyond them to the implications for planet Earth of a global economy of over ten billion trading consumers. Barring cata-strophic events that may well derail its arrival, this defines the world order in the making. It is the present and cementing world order built upon nonhuman genocide and ecological impoverishment.

From Biosphere to Homosphere

It is not all life that humanity's totalitarian reach is endangering, but the diversity of life at the levels of species, subspecies, diverse popula-tions, ecologically unique places, biological phenomena like migrations and other wildlife behaviors, and the evolutionary potential of much complex life. In the midst of this crisis, there are and will be species who survive and even prosper: they tend to be the hardy generalists, typically swift breeders, and capable of adapting to a range of habitats, climate envelopes, and foods. The Norwegian rat is a poster case of a generalist species and an admirable creature in its own right. The num-bers and ranges of ticks in parts of North America are also expanding, favored by the rise of Earth's average surface temperature.

Life-forms who are large and have extensive spatial requirements, are specialized in terms of food or climate needs, cannot move swiftly or cannot find places to move to, are adapted to small ranges (as in the tropics, islands, rivers, lakes, or mountain slopes), procreate slowly, are collected because people find them fanciful or profitable (like parrots, tortoises, seahorses, and orchids), and are killed legally or illegally for any number of reasons—these are the vanishing ones. Simultaneously and relatedly, humanity continues to appropriate enormous terrestrial and marine areas, erasing their rich ecological heritage and convert-ing them into biologically depauperate places under the yoke of human colonialism.

The destruction of life's intrinsic variety and abundance profoundly erodes the experience—human and nonhuman—of life's dynamic expressions and complexities. As the living world is diminished, hu-man ignorance of the full spectrum of the biosphere's magnificence deepens. Landscapes and seascapes lose the life-forms that composed and adorned them, becoming bereft of dimensions of intricacy, unexpectedness, richness, and uniqueness. The consequence is the downgrading of the biosphere's wealth and the banalization of places. The experiential upshot is the dissolution or diminution of beauty. The cocreation of the biosphere's dynamic canvas by innumerable life-forms comes undone. This dismantlement defines the biophysical tyranny of the human over the world: Just as political tyranny works by terrorizing a populace into silence, so human tyranny works by silencing nature through shuttering dimension after dimension of life's songs and works.

"Human beings are rapidly becoming a monoculture," states zoologist Aubrey Manning. "We suck resources in at the cost of the rest of life on the planet."[174] "As native species retreat and disappear to be replaced by alien competitors from other lands," Wilson writes, "global biodiversity is declining and with it differences in life-forms from one place to another."[175] "Under business as usual," state biologists Paul Ehrlich and Robert Pringle, "biogeography will become increasingly homogeneous."[176] Director of the Center for Biological Diversity Kieran Suckling points out that (present trends continuing) the future geological record will display the disappearance of biodiversity alongside the mass proliferation of domestic species (like sheep, pigs, chickens, cattle, wheat, and rice), revealing Earth's biological homogenization in our time. "A name along the lines of the Homogenocene," Suckling adds, "is fitting for such a period."[177]

What does the Homogenocene epoch bode for human beings? Could humanity survive a mass extinction of species and the precipitous impoverishment of life? This is a question that arises regularly in the environmental literature, and the truth is no one really knows the answer. Some scientists warn that humanity and civilization are imperiled by the loss of the "human life-support systems" of diverse species, healthy populations, and wild ecosystems, while others contend that humanity might do just fine on a human-dominated, hotter planet bereft of wild forests and wild fish.[178]

The enigma of our species' survival is arguably a distraction from a host of burning issues that life's collapse presents. It seems possible, after all, that humanity could survive profound losses of life and go on to occupy a planet destitute of biodiversity and awash in constructed environments, domestic animals, croplands, and generalist species, with scattered relics of wild areas here and there. Given such a bleak scenario (within which *Homo sapiens* indeed survives), doesn't the question of survival seem somewhat off topic? When we consider what human survival looks like after the Sixth Extinction, wouldn't we prefer—wouldn't we choose in a heartbeat—to change historical course so as to preserve Earth's rich community of life?

Matters of paramount importance press on human beings collectively at this historical juncture. The most outstanding is not how long we remain in existence (an unanswerable question), but who we are while we exist. How we treat our home planet—home to millions of species and billions of living beings—is foundational to human identity, and foregrounds questions that cannot be forever evaded: Will humanity settle for being a planetary colonizer with little to no regard for the intrinsic being of nonhuman life? Capturing the same existential and ethical aporia from another angle, Wilson asks: "What kind of an entity are we to treat the rest of life so cheaply? What will future generations think of those now alive to make an irreversible decision of this magnitude so carelessly?"[179] Will humanity sleepwalk its way into the night of mass extinction without the faintest clue that the biosphere—when left free to be what it inherently is—generates abundant sources of livelihood, well-being, beauty, inspiration, and mystery to sustain human beings physically, enlarge knowledge horizons indefinitely, and elevate the human spirit? Will humanity not awaken soon enough to the crassness and falsehood of viewing this oasis in the cosmos as a collection of resources? And how long will we turn a blind eye to the fact that the age-old plagues of social injustice, oppression, and war *presuppose* regarding the biosphere as a repository of resources for conferring opulence and power?

Without a momentous shift in historical course, "we know where biodiversity will go from here," note Ehrlich and Pringle: "up in smoke; towards the poles and under water; into crops and livestock; onto the table and yet into more human biomass; into fuel tanks; into furniture, pet stores, and home remedies for impotence; out of the way of more cities and suburbs; into distant memory and history books."[180] The question that most deeply confronts us at this historical moment—as we continue destroying Earth's biological wealth despite understanding well where this is leading—is not whether we will survive life's unraveling, but who we are as this dismal event unfolds under our watch.

"Catastrophe," writes author Deborah Bird Rose, "inheres in the destructive unmaking of the world of life that has been making itself so beautifully for so long."[181] The catastrophe's source is not a mystery. In two words it is human expansionism. By the same token, the solution to the catastrophe is also straightforward: we must halt our expansionism and contract humanity's invasion and exploitation of the natural world. We are called to scale down the human presence by low-ering our numbers, changing our consumption and trading patterns, and limiting infrastructural sprawl. Simultaneously, we must pull back from the natural world, restoring vast portions of land and seas to the generative power of wild nature, and reinstating the controls of Earth's workings to the biosphere's full house of life. Neither scaling down nor pulling back will diminish the potential for a vibrant human civiliza-tion, but on the contrary, these choices will be the wellspring of a new human identity freed from the ignorance and greed that the reduction of Earth to human ownership underwrites.

Since biodiversity's devastation and an impending mass extinction are directly due to human expansionism, the question logically arises: Why is humanity not taking decisive steps to halt it? I argue that the worldview of human supremacy makes expansionism appear normal and allows humanity to stay on course.

TWO

Human Supremacy and the Roots of the Ecological Crisis

It's some kind of thing, it ain't us but yet it's in us. It's looking out through our eyeholes. . . . It puts us on like we put on our clothes. RUSSELL HOBAN

There is a sense in which human expansionism continues apace due to inertia, from habitual activities of billions and growing human numbers to the spread of consumer culture and multiplying nodes of economic connectivity. But a notable point is that most people (if they think of it at all) countenance the takeover of the planet with indifference or acceptance, and appear disabled from evaluating it as an aberrant phenomenon. This particular aspect of expansionism—the seeming normalcy of the planet's colonization—springs from the prevailing worldview of human supremacy or anthropocentrism.

I use "human supremacy" and "anthropocentrism" as synonyms, so as to clarify the latter vague and apolitical concept via the readily available virulent implications of the idea of "supremacy." The term "anthropocentrism" is often taken to denote the embrace of those values and activities surrounding the natural world that serve human interests.[1] I do not use the term in this sense, but, in the tradition of deep ecology, I regard anthropocentrism as a human superiority complex that represents and treats the more-than-human realm as inferior, usable, and expendable. In the words of author Patrick Curry, anthropocentrism refers to "the unjustified privileging of human beings at the expense of other forms of life."[2]

Flagging the interchangeability of "human supremacy" and "anthropocentrism" is not simply a semantic clarification, but intended to highlight a substantive point reiterated throughout this work: anthropocentrism does not in the least serve human interests—anymore than white supremacy has ever served the ostensible interests of the putative Caucasian race. All forms of supremacy entrench violence as a way of life, which, beyond the obvious grave harms it inflicts on the denigrated, profoundly disgraces the perpetrators themselves. To drive this point home through example, the horror of witnessing baby seals clubbed to death, dolphins viciously stabbed, or thousands of chickens tightly piled in battery cages gives rise, simultaneously, to empathy for the suffering of innocent beings and revulsion at the condition of the human debased to thug or callous user.

In simple terms, human supremacy can be defined as the pervasive belief that the human life-form is superior to all others and entitled to use them and their habitats. The core idea of the human-supremacist worldview is superiority, while entitlement describes how that idea is operationalized. This worldview is sometimes openly invoked to glorify the human race and sanction its modus operandi, but far more typically it works as an unconscious lens—and thus all the more profound an obstacle—that debilitates human beings from being appalled at humanity's bloated presence and impact. Human supremacy is not formally or explicitly taught. It is indoctrinated into humans from a tender age, without time-out, hammered into the human mind by innumerable conditioning feats of the dominant anthropocentric culture: a culture that does not simply include dimensions of domination over the natural world, but is entirely built upon and constituted through domination.

How human supremacy is conditioned into human beings deserves some unpacking. The sociological and anthropological terms "socialization" and "enculturation" refer to the molding of the relatively amorphous nature of a newborn human into a person who will be able to partake, more or less competently, in the sociocultural world he or she is born into. "Conditioning" is a kind of parallel process, transpiring alongside primary socialization most especially after an individual has acquired language. Conditioning installs—from early on and fairly securely—a cultural belief system about the fundamental nature of the world and the place of the human within it. Conditioning is thus a deep phenomenon, which arguably imbricates into a neurological substratum, for it teaches people not only how to comprehend the world's "natural order" but also how to *see* it. Conditioning thus produces what critical theorist Herbert Marcuse called a second nature of man.[3]

Western culture, in particular, conditions the social body into the worldview of human supremacy by indoctrinating, in a wide range of overt and subtle ways, that the human is special, different, superior, privileged, and (in our time) supreme authority and presence on planet Earth. Anthropocentrism, as has been noted, is not strictly a Western civilization phenomenon.[4] This work's focus on Western culture is warranted for two reasons: one, the West has arguably developed the most robust and historically sustained expression of human supremacy; and two, the West has today become the dominant socioeconomic civilization, infecting the entire globe with its particular strand of anthropocentrism.

People are inducted into all manner of social hierarchies, but none is more fundamental nor more "obvious" to the social body than the hierarchy between man, on one hand, and the rest of nature, on the other. This hierarchy has solidified as a shared perception and understanding; as a raft of widely deployed resource-laden concepts about nonhuman nature; and as an unremitting instrumentalism that is overrunning the natural world with virtually no restraint. It is a constructed hierarchy inculcated by means of "a socially engineered arrest of consciousness"[5] that quashes the sense of wonder toward nature and of astonishment toward existence itself that bubble artlessly out of the human.[6] The spontaneous feeling of awe is suppressed, or trivialized into clichés, so that the natural world can be turned into, and accepted as, a human colony: a domain fundamentally for using and exploiting and which includes many nonhumans who are, by fiat, enslavable and killable.

What makes human supremacy a worldview is the enormous ground it covers in defining the human relationship with Earth's beings and places and with Earth as a whole. What makes the worldview so powerful is that it remains tacit: people rarely if ever explicitly entertain the propositions "we are superior to all other life-forms" and "we are entitled to use them." Instead, human behavior complies with those statements—which is what makes them assumptions. Because assumptions are only dimly perceived, if not completely undetected, they are difficult to dislodge from the human mind by comparison to overtly held beliefs that can be subjected to critical thinking. The dominant culture's assumptions structure "the context in which we think," while themselves remaining "almost inaccessible."[7]

The human-supremacist worldview is the deepest causal layer of the biosphere's plight, for it makes humanity's expansionism appear acceptable and inevitable. In our time, human supremacy—the shared belief that humans are above everything and can rightfully use it all—sustains the trajectory of history's course. The received credo of superiority-cum-entitlement reassures the human mind that colonizing the planet is the prerogative of our species' distinction: if not our manifest destiny, then our naturally ordained lot.

The way the belief system of human supremacy triumphs is also more subtle. Every imposition on nonhumans and the natural world appears compartmentalized in its own domain, seemingly isolated from everything else, and entraining its own benefits, costs, quandaries, practices, need for reforms, or what have you. So, for example, industrial fishing, sows in gestation crates, burning rain forests for soybean plantations, building roads and laying pipelines wherever, mining the seabed, locking up wild animals in zoos and aquariums, escalating industrial aquaculture operations, damming rivers, calling wild animals "game" or "trophies" and domestic animals "livestock," and so on, all such practical and conceptual ventures appear disjointed—as though (more or less) unrelated to one another.

It is only when we consider such impositions comprehensively, and especially when we tune in to the arrogance and violence that links them, that we become aware that they are all so many offshoots of the human-supremacist worldview. Because of that worldview, all such impositions are made possible. Within that worldview, they are all "one taste."

The worldview of human supremacy ramifies ad infinitum in applications and extensions. For example, human beings can go anywhere they choose, construct as many roads in any places they desire, make animals perform for entertainment, or maximally crowd chickens in order to maximally mass-produce cheap eggs and meat. Humanity gives itself permission to convert entire biomes, like grasslands, forests, wetlands, and coastal seas, for corn, wheat, meat, cotton, palm oil, soybeans, and salmon and shrimp farms. It gives itself license to plunder continental shelves and seamounts. The actions human supremacy sponsors cover the gamut from the seemingly trivial to the gigantic: if wasps build a nest on one's deck, one can purchase a lethal chemical and spray them; if there are coal seams in the mountain, companies can blow off the mountaintop, dump it over the edge, and get the coal.

Seeing how the supremacist worldview underlies such a span and number of actions is precisely seeing *that* it is a worldview. A worldview is a cultural-interpretive system that implies far more than a set of reigning ideas: it constitutes a lived belief system within which values, ideas, assumptions, and actions are intertwined; it spawns certain ways of thinking and being in the world, while precluding others; it binds "consciousness more or less blindly to inherited interpretations and does not permit consciousness of the possibility of alternative inter-pretations to arise."[8] The tremendous scope of the human-supremacist worldview bestows it with near omnipotence; it's as if the whole world is dyed in it. Indeed, this worldview is physically entangled with the biosphere because, to a large degree, it has sculpted the biosphere. Hu-man beings live in full immersion within it, making it difficult or even impossible for them to see it. By the same token, however, the world-view's colossal scope is its Achilles' heel: once exposed, it has nowhere to hide; once one sees it, there is nowhere it can't be seen. Unmasking the operative assumptions of superiority and entitlement enables the human mind to discern the full gamut of human-supremacist tyranny: the same vulgar exercise of power in the chemical asphyxiation of a wasp family and the demolition of a mountain; in the clear-cutting of a forest and the gutting of a seabed; in the nonstop persecution of wild animals and the cruelty of factory farms.

Human supremacy elevates the human, while simultaneously eras-ing the perception of the nonhuman realm as self-existing and replac-ing it with a simulacrum of a realm that exists for human using. It is thus the very coinage of our conditioned brain that undergirds and permits the onslaught on the biosphere. Every imposition of human expansionism does not appear as an imposition, but, within a world construed as rightfully possessed, presents itself instead as more or less blameless and certainly customary.

This emperor has clothes that have been handed down and re-embroidered over the course of millennia. We need to strip off the clothes, so as to see nakedly what we are doing and where we are head-ing. We are unraveling the living world in tacit accordance with the received credo that humans are a superior race and that this corner of the universe is humanity's property. But we can leave this daydream behind, and wake up to the glory that is evanescing.

Natural-Born Colonizers?

In endeavoring to understand humanity's far-reaching impact on the biosphere, there exists a widespread propensity to view it as stemming from the nature of our species. Indeed, certain circulating storylines, in the environmental literature and the broader culture, locate the fount of ecological depredation in "human nature."

Ironically, the classic storylines about "human nature," invoked by different parties to explain the ecological predicament, clash as tempestuously as Scylla and Charybdis. "I know of no study which is so utterly saddening as that of the evolution of humanity," lamented preeminent nineteenth-century scientist T. H. Huxley. "Man emerges with the marks of his lowly origin strong upon him. He is a brute, only more intelligent than the other brutes, a blind prey to impulses, which as often as not lead him to destruction; a victim to endless illusions, which make his mental existence a terror and a burden, and fill his physical life with barren toil and battle."[9] Huxley's mentor, none other than Charles Darwin, held a somewhat different opinion: "Man," Darwin gushed, "may be excused for feeling some pride at having risen, though not through his own exertions, to the very summit of the organic scale; and the fact of having thus risen, instead of having been aboriginally placed there, may give him hope for a still higher destiny in the distant future."[10] Along this fault line, so poignantly articulated by friends and colleagues Huxley and Darwin, one stock circulating view sees humans as prone to selfishness, shortsightedness, greed, and aggression, and thus inclined toward nature plunder and power struggles for the procurement of resources; while another circulating perspective sizes up our species as endowed with such exceptional intelligence and culturally transmitted technological knack that our very makeup set us up to become a biogeological force on the planet.

Whatever the specifics of the "human nature" narratives in relation to the state of the planet, the bottom line is a proclivity to explain nature's takeover by our innate constitution, and, even more insidiously, make humanity's overreach continuous with the natural order. This perspective is often casually rehearsed as though self-evidently true. For example, an *Economist* special issue on the biodiversity crisis offhandedly reported that "in a sense, this orgy of destruction [is] natural. In the wild, different species compete for resources, and man proved a highly successful competitor."[11] In the wild, sometimes organisms "compete for resources," while other times they play, loaf around, cooperate, communicate, eat, mate, hibernate, travel, sing and chatter, or just enjoy being alive. At a more foundational level, in the wild, the creation of life's stupendous diversity is life's strategy for *avoiding* competition: "more living beings can be supported on the same area, the more they diverge in structure, habits, and constitution."[12] Sweeping pronouncements about the naturalness of the human "orgy of destruction," and about the intractable competitiveness of nonhuman and human life, should be warily eyed as ideologically inflected and unlearned projections onto the nature of "nature."

Deconstructing the worldview of human supremacy is a sounder line of inquiry for understanding the biosphere's plight and discerning how to chart a new historical course. Shining the spotlight on this ruling worldview rescues us from the fallacies of ideas about human nature as driver of the crisis. The fallacy of the "humans are selfish and brutal" view is an implicit invitation not only to reduce human nature to a one-dimensional fabrication, but to resign ourselves to human belligerence taking its inevitable toll on the biosphere. The fallacy of the "humans are exceptional" perspective is a summons not only to wallow in its smugness, but to embrace humanity's status as Earth's aristocracy and welcome human domination as a kind of biological providence. Let us tack sail swiftly and safely away from these colliding storylines of "who we are." For the gravest fallacy of *any* narrative that naturalizes the human impact is that it disables people from reflecting upon the inculcated beliefs of human superiority and entitlement—let alone perceiving how blinding and violence fostering these beliefs are.

Dissecting human supremacy shifts discussion away from different readings of our "species-being" to the decisive role played by civilization's shaping of human identity. Anthropocentrism is precisely the sociocultural molding of the human into an identity that deems itself above all other life-forms and at liberty to use them. This conditioned identity is not who human beings are by nature nor forever saddled with. The particular identity that presumes to hold a first-class ticket on Earth is not inborn to *Homo sapiens* but more analogous to a grandiose personality that humans are capable of donning, and apparently highly susceptible to given its cross-cultural contagiousness.[13] The ingrained conviction of human preeminence binds human consciousness tenaciously to a received way of life and blocks an alternative way of being from emerging.

The cultural installation of the beliefs of distinction and prerogative downloads the core credo of anthropocentrism. The programming of that credo into the human mind is accomplished through an indefinite number of ways as a lifelong conditioning. In the process of daily life, children learn by osmosis that environments are human dominated or legitimately subject to human domination. Later, through encountering geographical maps, and through education in political geography, human beings learn that the world belongs to people and has been territorially carved among nations. Formal schooling further teaches youngsters that history consists of a parade of civilizations, kingdoms, periods, wars, emigrations, treaties, grand monuments, technological inventions, and so forth—all the while transmitting the sublimated message that Earth is just the background upon which history unfolds. Ecological realities, contexts, and contributions are rarely if ever taught in history. In fact, the field of environmental history only emerged toward the end of the twentieth century.

Simultaneously with all this, human beings are taught the category of "animals" as a distinct class of beings from "the human"; moreover, many animals are understood, from an early age, to be property, vermin, game, or food—in other words, subject to absolute power. By the time a person formally learns that he or she also belongs to the kingdom Animalia, this knowledge is guaranteed to remain superficial, cerebral, and mostly irrelevant. Additionally, as the work of author Richard Louv has documented, human isolation from the more-than-human world is deepening in our time, fostering a kind of de facto human-centeredness.[14] Another way worth mentioning in which human supremacy is inculcated is through the teachings of traditional humanism: these extol the great achievements of literature, architecture, art, music, science, technology, and medicine with nary a nod—not to say expression of thanks—to the indispensable contributions of the natural world and of nonhumans to those achievements. Additionally, while traditional humanism has commended the virtues of community and solidarity between people, it has never extended those virtues to include the nonhuman world.

In a world dominated by Western civilization, industrialization, and domestic animals, human supremacy has come to manifest as three invisible shared beliefs: that Earth belongs to humanity; that the planet consists of resources for the betterment of people; and that human beings are of distinguished stature by comparison to all other species. These beliefs are "invisible" not because they are hidden or difficult to detect, but in the sense that they are rarely explicitly stated or reflected upon. They are clearly operative as unvoiced assumptions according to which people, corporations, nation-states, and other social entities comport themselves in the biosphere. The invisible operation of this belief system is the upshot of a long history of the self-positioning of Anthropos front and center: an unbroken tradition of anthropocentrism that reaches back to classical antiquity and has its roots in the birth of civilization.

The Origins of Human Supremacy

With the emergence of civilization some eight thousand years ago came the first large-scale "geoengineering" projects, including the clearing of forests, grasslands, and other natural vegetation to make way for crop agriculture, as well as the diversion of rivers into canals and impoundments for irrigation. Landscapes were modified and subdued for domestic grazing, while wild animals were systematically killed and driven into remote areas in order to protect the flocks. Alongside these developments emerged the first permanent human settlements, initially as small villages and later as more densely populated city-states, which were often encircled by defensive walls. The combination of sedentary communities, the beginnings of human population growth, the spread of domesticated animals and crops, and habitat takeover and conversion triggered a widening wave of ecological losses and extirpations of wild plants and animals.[15]

The ability to convert and control environments in hitherto unprecedented ways, preconditioned human consciousness favorably toward notions of human uniqueness. What's more, even at the most basic perceptual and experiential level of inhabiting semi-urban settlements, insulated from wild nature by cultivated and pastoral lands and sometimes actual walls, human consciousness became increasingly biased toward seeing the human realm as set apart from the more-than-human world. Over time, the environs became more and more human dominated, and wild nature was pushed further and further beyond the pale of encounter. With these sociohistorical and environmental changes, which unfurled gradually, a lived sense of separateness made the human mind fertile ground for essentialist ideas of human difference to take hold. As philosopher Boria Sax notes, at some point the division between humanity and nature ceased being strictly geographical and became folded into philosophical thought as well.[16]

It is beyond the scope of this work to detail in full the origins of anthropocentrism. What is clear, however, is that human-centeredness co-emerged with civilization—and especially through its herald of crop and animal agriculture. Indeed, the plow is counted as one of the "greatest inventions of all time," for allowing "agriculture to spread across fertile flat lands and push wolves, bears, tigers, and other wild beasts out to the wild and woolliest fringe places of the world."[17] By overtaking and repurposing entire biomes, agriculture undergirded the expansion of human settlements and numbers, which led to the increasing conversion of wilderness (including forests, grasslands, and free-flowing rivers) for cultivation, which in turn further augmented human settlements and numbers.

Anthropocentrism solidified over time through the interplay of a newfound, profound capacity to manipulate nature, on the one hand, and of musings about human distinction that soon gained cultural ascendancy, on the other. Around the Near East, the Levant, and the entire Mediterranean basin—the cradles of Western civilization—over the course of centuries and millennia landscapes became ever more humanized by settlements, farming and grazing, trade routes, and the parallel receding of natural areas, big carnivores, herbivores, and other animals. Civilized human beings began to nonconsciously condition themselves into a center-periphery relation with the more-than-human world, which over time eventuated into a virtual loss of sight of any other possibility of relationship. "With the transition to settled agriculture the relationship to the natural world and to wilderness changed dramatically," writes ecological economist Lisi Krall. "People lived in a world mostly of their making fostering a duality that had not been present for pre-agricultural people."[18]

Gradually human consciousness became molded in alignment with a way of life that rested on engineering landscapes and combating wild nature. Anthropocentrism—the human perception of living within an insular "self-referential system,"[19] of being in charge (relatively speaking), and of pride in being a distinguished life-form—must have overtaken human beings incrementally, reinforced over centuries by the human distance from and domination over wild nature. Foreshadowing the biblical story of human dominion, ancient Greek playwright Sophocles extolled man's overpowering of virtually everything (though he conceded the notable exception of death)—soil, land, birds, wild animals, beasts of burden, and elemental forces: "Many the wonders," was the verdict of the chorus in his *Antigone*, "but nothing more wondrous than man."[20]

Indeed, by the time the classical worlds of Greece and Rome emerged, the human mind was well seasoned for what political scientist John Rodman aptly called the "Differential Imperative": ideas of a human-nonhuman unbridgeable divide that philosophical, ethical, and political schools of thought would set forth.[21] To be sure, views championing the human difference were not the only ideas that vied for attention, but they became the ones that acquired premier status.[22]

Ideational and Physical Displacements of Disparaged Others

The worldview of human supremacy has operated by displacing non-humans, wilderness in general, and indigenous people, who were invariably labeled barbarians, savages ("colonialism's magic word"[23]), and primitive, or called beasts and wild animals. Human supremacy's displacements can be elucidated by grouping them into two comprehensive categories: ideational and physical. By ideational displacements, I refer to the dissemination of disparaging beliefs about nonhumans, wild nature, and native peoples—beliefs that enjoyed unbroken preeminence for millennia. Physical displacements refer to geographical conquests that have exterminated or dislocated multitudes of disparaged others, beginning with the agricultural way of life several millennia ago and continuing into our time.

The ideational displacements have been realized through a dominant Western schema that has inquired into "the human phenomenon" by posing the ubiquitous question: How are humans different? Rodman called this staple aporia of Western thought the Differential Imperative, writing that "the basic concept upon which the whole edifice of classical thought was built—the concept of human virtue or excellence—was defined by isolating the distinguishing characteristics of the human species from those of other forms of being, especially the brute beasts, our next of kin." Human distinction acquired "axiomatic status," he further noted, given "the almost universal tendency of mainstream classical writers, both pagan and Christian, to assume the Differential Imperative as self-evident."[24]

Over the long course of history, there have been no shortage of submitted distinctive characteristics of the human: Reason, language, morality, religion, soul, culture, technology, perfectibility, free will, and personhood have all been proposed as exclusive human qualities. These have been the most rehearsed attributes, but other distinguishing human features have also been entertained, such as erect posture, laughter, and private property. Some thinkers enjoyed stringing numerous unique attributes for the purpose of exalting the human: "To you is given a body more graceful than other animals, to you power of apt and various movements, to you most sharp and delicate senses, to you wit, reason, memory, like an immortal god."[25] Others preferred to rhapsodize over the scope of man's dominion: "The immense magnificence of our soul may manifestly be seen from this: man will not be satisfied with the empire of this world, if, having conquered this one, he learns that there remains another world which he has not yet subjugated. Thus man wishes no superior and no equal and will not permit anything to be left out and excluded from his rule."[26]

Differential Imperative ideas tended to gravitate around the human-animal contrast, and for good reason: since animals are clearly our closest kin, success in elevating the human into a standalone category called for the construction of a sharp boundary between "us" and "them." The blanket schema that "animality" is all about instinct, corporeality, stereotypical behavior, and biological evolution, while "humanity" is all about learning, reason, meaningful action, history, and culture, has, amazingly, governed Western thought from antiquity into recent times. This framework has worked to blind the human collective to the magnitude, complexities, and wonders of nonhuman consciousness. This is indeed how sweeping frameworks trick the mind: they present themselves as windows onto the world, when in actuality they are portraits that cover up or distort reality. As author Matthew Calarco clarifies, "the classic human/animal distinction serves to block access to seeing the world from the perspective of nonhuman others and seeks to limit in advance the potentiality of the animal and entire nonhuman world."[27]

Philosopher Giorgio Agamben has argued that the human-animal divide is the founding gesture of Western civilization.[28] Indeed, the human-animal divide is the central pillar of the supercilious human identity that civilization has forged. As I argue in detail elsewhere,[29] establishing this divide has been a core strategy for the domination of the natural world more broadly. For by stabilizing human ascendancy over "the animal," a vast ontological distance between humans and everything else (like plants, mushrooms, or rivers, for example) has been automatically guaranteed; by constituting "the animal" as inferior, everything else could easily be regarded as even lesser or merely physical; and by denying "the animal" moral consideration, the moral standing of the entire natural world was automatically rendered superfluous. What's more, as Jacques Derrida elaborated, lumping a vast diversity of beings into the linguistically homogenized category of "animal" has served as an ingenious conceptual move for singling out and exalting the human.[30]

Whatever the favorite philosophical, theological, political, or other Rubicon, "the search for [an] elusive attribute" of human uniqueness and superiority "has been one of the favorite pursuits" of Western thinkers.[31] What the various distinctive qualities share is the assumption of a definitive polarity between humans and nonhumans. As one popular eighteenth-century English writer pithily summed this ostensibly clear-cut division, the line between man and the rest of nature is "strongly drawn, well-marked, and unpassable."[32]

The human distinctions flowing from "Western Civilization 101," so to speak, have primarily, and certainly as a distilled conditioning missive, not only exalted Anthropos and his supposed specialness but simultaneously portrayed nonhumans (and for a long time "inferior" humans) as deficient by comparison. The quest for human distinction also functioned as the cornerstone trope for the elaboration of hierarchical narratives. The most enduring of these—historically threading across very different traditions of thought—has been the Great Chain of Being: this grand narrative ordered Creation as a graded hierarchy from pure spirit to inert matter.[33] Within the Great Chain humans were positioned at the apex, just beneath angelic beings and God, while animals, plants, and minerals followed down the line.

This prevailing model in Western history was cognitively appealing for organizing Creation in a tidy order; and it was sociopsychologically appealing for giving humans pride of place. Within the Great Chain of Being, each domain was said to rightfully use the one beneath it; for example, animals were entitled to use plants and plants to use minerals. Since humans occupied the highest earthly rung, they were duly authorized to use all other beings and domains.[34] Thus the Great Chain has not only functioned as a complete description of Creation (what philosophers call an ontology), it has also worked as a moral order sanctioning the use of everything. The achievement of the Great Chain of Being was to fold the beliefs of human superiority and entitlement into a single cosmological package. It is perhaps not surprising that this ontological-moral order has endured for so long: it is immediately accessible to everyone—from the most educated to the completely illiterate—and it is serviceable in giving license to everyone to have one's way with nonhuman nature.

Working in close formation with ideational constructs of others as unworthy of respect or consideration have been the physical displacements of the more-than-(civilized)-human world, effected primarily by means of conquest. The modality of conquest eschews what philosopher Martin Buber called an "I-Thou" relationship for an "I-It" connection.[35] The human occupation of geographic space is prototypical of an I-It arrangement. At will, the "I" (human beings enculturated into a supremacist worldview) owns, appropriates, and repatterns everything—from above ground, underground, in the seas, or beneath the seas. (Technology permitting, this mode of operation will eventually also be transferred to extraterrestrial bodies, as we are beginning to see.) The nonhuman life killed or displaced in the process of conquering geographical space has been dignified with neither attention nor compassion.

We inhabit a world in which a series of civilizations, empires, and societies have destroyed forests, plowed grasslands and shrublands under, overgrazed landscapes, drained wetlands, dewatered and diverted rivers, exploited lakes and seas, overfished and polluted the ocean, and roundly used everything. Alongside exploiting geographical regions, wild animals have been killed, persecuted, enslaved, forced to flee to remote regions, and driven to functional, regional, or total extinction. Moreover, as historian Keith Thomas noted, "the idea of human ascendance has implications for men's relations to each other, no less than for their treatment of the natural world."[36] Thus, so-called inferior humans were called beasts, savages, and the like, and have endured humiliations, subjugations, and genocides. For example, at the turn of the eighteenth century, European colonists hunted down the Indians of New England with dogs; "they act like wolves and are to be dealt withal as wolves," proclaimed a clergyman as justification.[37] The human-supremacist worldview has never placed all humans on a par, but deemed that only certain humans truly embodied the qualifying distinctions of the superior prototype. As Calarco argues, anthropocentrism has historically included "only a select subset of human beings . . . within the sphere of humanity proper."[38]

Briefly put, self-willed nature, wild animals and plants, as well as so-called uncivilized peoples have been subjected to nonstop incremental and large-scale physical dislocations. The biosphere has been biophysically sculpted by a human-supremacist orientation that has driven such dislocations and inflated itself through them.

The Worldview of Human Supremacy Shapes the World

Over history's course, the mutual interplay of *ideas* vaunting human difference and *acts* of geographical conquest have empowered and consolidated anthropocentrism. These dimensions have worked synergistically. The nonstop physical expansion of civilized humans yielded the shared experience of increasingly tame landscapes and of human power over them, thereby biasing the collective consciousness toward embracing the conviction of supremacy. At the same time, Differential Imperative ideas—portraying the human as superlatively endowed and the nonhuman as patently lacking—justified expansionism. With the conquest of landscapes and sea routes burgeoning, as early as late antiquity and the dawn of the Christian era the known world came to be called the ecumene, meaning "the inhabited world as a whole, as the common possession of civilized mankind."[39] The invention of human specialness, alongside the emergent anthropocentric meme of human planetary ownership (ecumene), began to secure its mesmerizing grip over the social imagination.

Far more than a representation of how things putatively are, human supremacy is a worldview that has forged the world humanity inhabits, both mentally and physically. The ideational and physical dimensions of anthropocentrism have reinforced one another in the pattern of a positive feedback loop. Physical power over the nonhuman world has buttressed the solidifying consensus of human preeminence; and that prevailing conviction has fueled the continual takeover of "inferior" realms and beings. This snug interplay of cognitive belittlement and physical conquest is the defining dialectic of human supremacy. The synthesis of its allied ideological and material forces of domination precisely constitutes its solidification into a worldview: one that has been historically bequeathed and reaffirmed (with some modifications) without interruption, across major socioeconomic regime changes, as well as across the big divide from a religious-traditional to a secular-modern social order.

Human supremacy has prescribed an actionable ideology for setting upon the world. Ultimately, this worldview has sponsored the occupation of the biosphere. The corralling of wild nonhumans, indigenous peoples, and wild nature into the geographical margins (such as reserves and reservations), often outside the preoccupations of collective consciousness, and the simultaneous humanization of the world—the globalized ecumene—have yielded a widely shared mental schema of Earth: Earth increasingly appears as a physical backdrop (and even a "starter-planet") for humanity's unfolding destiny and a stage for civilization's continued march.

The original ontology of Earth as something greater, inexhaustible, unknown, enchanted, and mysterious has been supervened, while a man-made ontology of the civilized human has become physically entrenched and conceptually reified. The vast reality of Earth as a field of multitudinous beings and relationships, as a seemingly indomitable domain, as a realm harboring the magical and the unexpected, as a living canvas of experience and meaning far wider than the human sphere (and encompassing of it) has dwindled. In the place of that vast reality a humanized reality looms. This ontological inversion of a part (the human) claiming the whole (the biosphere) has transpired over history's long course. Indeed, once the ideational and physical jugger-naut of human supremacy was set into motion several millennia ago, it was only a matter of time before man and his works, to paraphrase social theorist Guy Debord, would become virtually all there is to see.[40]

Human supremacy has driven Earth's terraforming on such a totali-tarian scale as to constantly reinforce and reactivate its own premises. We are the inheritors of this worldview's long historical march. We are, more starkly stated, the *products* of that history. Humanity's way of re-lating and impact on the natural world is rooted far more deeply in the historical legacy of anthropocentrism, than in either fiendish flaws or superlative attributes of human nature.

Human Supremacy Blindsides Humanity

From classical antiquity through the present, thinkers overwhelm-ingly chose to focus their intellectual energies on assertions about the human difference. That choice, alongside the displacements such as-sertions validated, became implicated in engendering and sustaining a worldview that has not only devastated those deemed unworthy of respect (beings and places), but also been disastrous for the ostensibly distinguished humans themselves. "Human self-enclosure," to borrow environmental philosopher Val Plumwood's description of human-centeredness,[41] has generated tragic blind spots in humanity's rela-tionship with the biosphere and promoted human conceit, which—in sharp contrast with the grandiose intent of the supremacist credo—has debased human dignity and perhaps fatally undermined the potential for a beautiful human inhabitation on Earth.

As discussed so far, anthropocentrism's rule can be dissected in terms of who and what have been exterminated or dislocated beyond collective awareness. Another way to scrutinize human-centeredness—bringing us from a different angle to the same outcome of the biosphere under siege—is through posing the question: What has the supremacist worldview done to *human beings* who come under its spell? I examine two momentous consequences: first, this worldview has yielded a col-lective incapacity to stop or limit human expansionism; and second, it has blinded humanity to the profound loss of the biosphere's richness and beauty and to the conventionalized violence that characterizes hu-manity's dominant way of life.

The wisdom of limitations belongs to human cultures and individuals who respect their nonhuman (and human) neighbors. Respect naturally gives rise to restraint.[42] The human-supremacist worldview, however, extinguished respect for nonhuman neighbors and neighborhoods via circulating demeaning beliefs about them: that they do not morally count or are inferior or are nothing but resources. The conditioning of the human into self-acclaimed ascendancy precludes the arising of restraint that flows from respect, or from a sense of awe toward world and existence, which would generate the desire to fit in with the more-than-human realm rather than convert and displace it. Nonstop expansionism is thus built into anthropocentrism, for inculcation into the human-supremacist credo removes barriers to plunder. What's more, forests, rivers, mountains, wild animals, and indigenous people have been unable to thwart the plunder. Resistance against the civilized conqueror has indeed been futile.

From the perspective of civilized humans who have deemed their stature unparalleled, there has never been any ground from which to discern, or be forced to discern, reasons for restraint. Limitless expansion has been a presumed and exercised right for millennia. This is a straightforward upshot of the anthropocentric worldview. The atrophying of wisdom that is affiliated with restraint has thus *happened* to humanity beyond deliberate choosing, and is among the indignities that civilized humans have unwittingly reaped.

Human supremacy has spawned an enterprise that only knows how to grow—to invade and assimilate, to convert and develop, to acquire and consume. Indeed, the emergence of the religion of growth in the modern era has its historical roots in the anthropocentric worldview. "Growth" was just the celebratory label that human supremacy gave to its mode of operation after it became self-conscious of the seeming triumph of its reign. But the modern celebration of growth represents nothing but the coming of age of a stunted human imagination, one incapable of envisioning a different way of being on Earth other than the one that the history of human supremacy has fostered. The wisdom of limitations has been rendered so inoperative in our time that the mainstream is virtually tone-deaf to eleventh-hour appeals for it.

The absence of restraint today has become a runaway syndrome. We see it in the atrocity of factory farms; the trashing of the seas and wholesale destruction of marine life; the drive to dam the world's remaining free rivers; the rendition of whole landscapes in pursuit of natural gas; the mad scramble for offshore oil; the trade-off of irreplaceable rain forests for plantations and ranches. We see the inability to countenance restraint in the acquiescence to the march of our numbers upward of ten billion. We also see the inability to countenance restraint in the defiance of the number 350 parts per million: civilization seems unable to halt its expansionism even in the face of civilization-endangering climatic chaos.[43]

The cultural and institutional accretions of the long history of human supremacy have rendered dominant civilization incapable of observing limitations for the sake of the beautiful cadre of life that remains with us. In our time, the mainstream human enterprise is seeking to demonstrate that our expansionism can continue—as though humanity's sojourn on the planet were a game of Russian roulette and human beings might still turn out to be the fabulously rich winners.

The inability to see any point in limiting the human enterprise is one fatal blind spot of anthropocentrism. A second is that anthropocentrism has blinded humanity to the destruction of the biosphere's magnificence. An insight in the work of social theorists Max Horkheimer and Theodor Adorno speaks eloquently to this point. "Men pay for the increase in power with alienation from that over which they exercise power," they wrote.[44] The price of human supremacy has been exactly that: as civilized humans' power over the natural world has grown, so, by the same token, has their blindness to the living wonders of the biosphere as well as to the violence they have unleashed within it. It is the most bitter of payments. Becoming alien on our home planet is humanity's ultimate reckoning for opting for the shallow privilege of power.

Perhaps nothing better displays human alienation from the natural world than "the shifting ecological baseline." "With each ensuing generation," biologist John Waldman explains, "environmental degradation generally increases, but each generation takes that degraded condition as the new normal." The passage of time everywhere reveals, in Waldman's words, "the insidious ebbing of the ecological and social relevancy of declining and disappearing species [and ecologies]."[45] "Animals, plants, habitats, and human cultures vanish," writes marine biologist Carl Safina. "Even the memories of them are disappearing."[46] A deepening collective oblivion of the primal richness of life, and of the obliteration of countless profound human experiences in the biosphere, accompany these losses. The downwardly shifting ecological baseline testifies to the intergenerational erasure and forgetting of beings and places. Erasures have occurred both incrementally and in quantum leaps, but forgetting has largely been swift and final.

Along these lines, we can discern that human supremacy manifests almost as a cult—the cult of humanity—wherein all human heritage (cultures, ancestors, nationalities, statesmen, architecture, and various civilizational achievements) are immensely valorized in traditions and practices of retention, restoration, remembrance, and tribute. The cult of humanity emerges out of the wedding of the artifactual restructuring of the world and the trapping of human attention within that restructuring. Everything else gets backgrounded. Biological heritage, in the sense of the natural environment with which cultures and humanity as a whole coevolved, has been deemed irrelevant, dispensable, moldable, and forgettable. Human supremacy thus encourages insularity and solipsism: as a lived worldview it has developed endless ways to use the more-than-human world, but no deep and enduring traditions of honoring it for what it intrinsically is and for who coexists with the human. The supremacist worldview has succeeded in suppressing the simple yet profound truth that nonhumans have "their own existence, their own character and potentialities, their own forms of excellence, their own integrity, their own grandeur."[47]

The declining ecological baseline is an existential condition that comes in the package of human self-enclosure, and it is linked with "the homelessness of contemporary man"—homelessness being the reckoning of arrogant oblivion to what is near the human, above, below, and all around.[48] The declining baseline does not reflect biological limitations of human memory or perceptual shortcomings, as some have argued.[49] It is a constituent feature of human supremacy: it stems from an absence of interest in and cultivated blindness to the intrinsic being of the nonhuman, which in turn translates into the absence of traditions within the dominant culture for recording and honoring the nonhuman world. This sociocultural ground of the declining ecological baseline serves to preempt any deep questioning; it thus also smothers potential grieving from arising. For example, the Yangtze River dolphin, called the baiji, was declared extinct in 2007. All the dolphins died, as their river became unlivable by noise, pollution, siltation, ship traffic, and overfishing. A remarkable footnote to this tragedy is that people who live near the dolphins' former home already do not remember the existence of the baiji.[50] Within the logic of a supremacist global culture this makes sense; if people did remember, they would question what was done to the river and would grieve for the place and its beings. At the same time, the collective "blackout" concerning the baiji's extinction (and existence) raises an eerie question: if people, conditioned into a human-supremacist culture, can forget a dolphin, what life-form or living place can they not forget?

Human self-enclosure obliterates mindfulness toward the living planet. Each generation regards the state of the world they encounter as the norm, while any collective sense of loss regarding the decline of the more-than-human world is absent (or dissed as a flaky sensibility of "tree-huggers"). Since wild beings have not been neighbors, nor wild places neighborhoods, neither their presence nor their disappearance has for the most part warranted recording. Indeed, scientists seeking to reconstruct historical abundances and former ranges of nonhuman species must engage in imaginative detective work and creative inference; straightforward historical-ecology information is rarely available.[51] Amnesia about the living world and its diverse beings is the wretched existential condition humanity has obtained in exchange for domination.

A pinnacle of alienation from the natural world in our time has been the perverse public invisibility of the mass extinction episode on the immediate horizon. Human-driven mass extinction remains publicly largely unknown, little understood, rarely talked about, or summarily glazed over with platitudes. Its invisibility is testimony to the incapacity of the supremacist mindset to acknowledge and face up to the consequences of its Earth treason.

The penalties of decreeing humanity as first and foremost have been the inability to limit the human enterprise's expansionism and a blindness to consequences that could cast doubt on human glorification. Thus have civilized humans continuously expanded and everything annihilated on their path been forgettable. "A Greek of noble descent," wrote philosopher Friedrich Nietzsche, "found such tremendous intermediary stages and such distance between his own height and that ultimate baseness [of the slave] that he could scarcely see the slave clearly."[52] Similarly, when man disparaged the more-than-human world as the beneath-the-human world, humanity largely forfeited the clarity to see, let alone grieve, the retreat of earthly marvels. The loss of marvels was abetted by a failure of sight that has always accompanied the imagined heights of Anthropos and his fabricated distance from all else.

In brief, human supremacy can be deconstructed from the lens of ecological justice: for its repercussions for nonhumans, native people, and the natural world; the ways it undergirds the takeover and conversion of wild nature; how it sponsors unnecessary suffering that is unperceived or ignored; and how it legitimates or tolerates actions that drive to untimely extinction and death countless life-forms. Yet it is not only the nonhuman world that suffers the violence of anthropocentrism. Supremacist humans are also victimized by this worldview: for it has brainwashed people in such a way as to desiccate the imagination for a different way of life, destroy the capacity to recognize the loss of wonders, and disable the discernment of everyday violence toward the nonhuman realm.

Questioning anthropocentrism is far from armchair philosophizing. It is an endeavor to move beyond problematizing the symptoms of this "age of crises," in order to wrestle with the conceptual and materialized foundation of the civilization that is causing them. What has it meant to elevate Anthropos and place him in the center? One way to answer that question is to observe who, in mirror image, has been marginalized and pushed to the periphery, and to take note of what voices have been silenced and who has been decreed to have no voice.

Responding to the question in terms of what humans have reaped from their self-elevation circles on matters of power and yields irony (or better, tragedy). Exalting the human has ostensibly been serviceable in acquiring power by means of configuring the world as a place to plug into, and source from, in whatever way called for. For example, biomass converted to food factories; rivers dewatered for agriculture; the Earth's crust scoured for energy; marine life defined as "fisheries"; rhino horn, tiger bone, bear bile, elephant tusk, pangolin scales, chiru pelt, shark fin, or snow leopard fur extracted by dark and vile means to serve empowerment fantasies. The human center always manifests as operationalized, unending entitlements. The center expands and the periphery recedes until the human sees only Anthropos, both literally in a humanized world and cognitively by losing stereoscopic vision of the other, losing inspiration for another way to live, losing the capacity for awe, losing the will to act from compassion, becoming blind to the ubiquity of violence. The spurious act of human self-elevation has betrayed the very source from which real power springs: from Earth's gifts of abundance, creativity, reciprocity, enchantment, unexpectedness, and belonging. The colonizer has thus built a trap for himself, as his pitch for power is now ricocheting into a powerlessness to honor limitations and an impotence to stop the violence. "Have I not reason to lament," cried out a poet, "what man has made of man?"

Human supremacy has diminished humanity profoundly, imprisoning human beings inside a hierarchical, crass, and ignorant worldview, and constricting their understanding and collective experience on Earth. "Man has closed himself up, till he sees all things thro' narrow chinks of his cavern," wrote William Blake. But "if the doors of perception were cleansed everything would appear to man as it is, infinite."[53] "Anthropocentrism" may sound like academic jargon, yet it's not. It is a lived worldview that has butchered the world and programmed our thinking and perception. But it ain't us.[54] What is the use, after all, of being distinguished and due special prerogatives when what that yields is the turning of Earth into a dilapidated, life-impoverished planet? Let's take off the clothes that have been putting us on, choose a planet of life, and join the world of the living.

The reductive claim that human beings make on the living world—in our time by means of its conceptual-pragmatic transmutation into a resource domain to be managed—lays a suffocating claim on humanity itself, trapping it in a cavern of contracted thought with no vista for a higher way of life.

THREE

The Framework of Resources and Techno-Managerialism

Specialists without spirit, sensualists without heart; this nullity imagines that it has attained a level of civilization never before achieved.   MAX WEBER

Human supremacy is violent—and often capable of heart-stopping violence.[1] Its violence, however, remains invisible by means of ontologically downgrading those upon whom violence is directed. The ontological downgrading is a historical attainment of the propagation of dismissive ideas about the nonhuman realm that originated as answers to the protracted obsession with the "human difference." This persistent line of inquiry eventually congealed into the Great Chain of Being, which gathered the totality of living beings into one comprehensive hierarchy. Forthwith, a myriad of belittling ideas about and brutal actions toward nonhuman life, like so many iron shavings, could adhere and fancy being in legitimate standing on the magnet of this grand narrative.[2]

Before turning to the spell of this hierarchical narrative that endures today, a caveat is in order. We live in a time where new discoveries and ideas, often clashing with the human-supremacist worldview, are jostling for attention, including the evolutionary kinship of all life; inquiry into animal minds, tool use, and language; and emerging studies of plant and fungi intelligence. People from many different walks of life are energetically contesting human empire. For example, Christian and Jewish voices, and recently Pope Francis, are challenging anthropocentric interpretation.[3] In short, our age is not unified by any one particular credo about either nature's order or the right relationship between the human and more-than-human realms.

While it may thus appear that in the modern age the cosmology of the Great Chain is no longer credible, nothing could be further from the truth. The Great Chain of Being with man at the top is everywhere: it is the "dark matter" permeating the world. Modernity did not abandon the Great Chain but gave it a new twist.

Hiding the Great Chain of Being inside Language

In modern times, the world-making narrative of human ascendancy and nonhuman demotion was secured in collective consciousness by being submerged into concepts that directly emerged out of the inherited worldview that the natural world is beneath and for humans. The most overarching of modern-age anthropocentric concepts is that of *natural resources*, which linguistically crystallized the post-Cartesian secular view of nature as purely material, mechanical, and lacking inherent purpose.[4] In the classical and Christian eras, Western thought championed the belief that animals, plants, minerals, and "savages" existed, or were created by God, for human use. This doctrine laid the sturdy foundation for the modern-secular concept of resources to emerge: Nature as a whole became a domain for human use and advancement. Indeed, the very category of "nature" co-emerged with its construction as precisely such a domain.[5]

The concept of resources not only foregrounds instrumentalism—the lens of viewing things as useable and profitable—but being general, content-less, and spiritless, it enables instrumentalism to engulf the entire reality field. Resources is a label that can be foisted on anything: soil, water, animals, minerals, rivers, forests, grasslands, and even extraterrestrial territories like the moon and other planetary bodies. It casts a blanket of inertness or gross materiality over the substance of the world, whose aliveness and self-integrity are denied or shunted into irrelevancy. This master concept legitimates taking and using, because it describes the world in ways precisely intended to invite taking and using. The instrumental reasoning and justification for domination inherent in "natural resources" attained altogether new heights through the industrial subjugation of soil, land, seas, animals, plants, and biomes by means of technologies invented and implemented without ethics. The amoral modality of industrial technology maps onto the concept of resources—for "resources" are constituted as lacking inherent standing. Amoral technology conforms to, as well as reinforces, the idea of natural resources because setting upon the world in accordance with that idea concretizes and legitimates it. People can thus countenance battery chickens, lopped-off mountains, endless monocultures, and industrial fishnets bursting with fish not as crimes against nature, but as the way things are.

The pervasive and customary use of "natural resources" makes it appear a normal way of talking about the world, as a language containing no agenda. It is worth scrutinizing what kind of relationship to the biosphere this ostensibly neutral concept fosters. The term "resources," simply put, designates the world in terms of its disposability for human needs, wants, and desires. "Natural resources," while seemingly an objective referent to (nonliving and living) entities, tacitly reconfigures the natural world as human owned. The concept entirely bypasses, and via its ceaseless use ends up erasing, the natural world's intrinsic standing. Indeed, the concept of resources preempts human thought from even moving in the direction of understanding nonhuman nature in terms of its intrinsic being. "Resources" prefigures and justifies the assault on the natural world. Nonstop exploitation is thus codependent with the omnipresent idea of natural resources—an idea that is beholden to, and a subterraneous continuation of, the historical legacy that nature's very purpose is to serve humanity. The transfiguration of nature into resources shapes human thought and action at such an all-encompassing level that people end up perceiving the biosphere through this single framework. An alternative way of relating with the nonhuman realm and another way of seeing it become erased or profoundly attenuated.

Nature's configuration as resources is a supremacy-laden concept with the power to abet anything—decapitating mountains, strip-mining seafloors, damming rivers, fracking land and seabed, building roads wherever, or piling domestic animals into neon-lit, sickening facilities and stuffing them with unnatural foods and substances. Liquidating others, be they places or beings, is built into the concept of resources and operationalized in practice.

Resources is an umbrella concept that provides a readymade, unassailable rationale for appropriating even that which has yet to be taken. For example, a panoply of technologies, planned actions, and legal arrangements are in place to seize the apple-sized polymetallic (or manganese) nodules, which exist, in apparently great numbers, on the ocean's floor. These "resources" contain manganese, nickel, copper, and other valuables: the only reason they have not yet been mined on a big scale is that they are not low-hanging fruit—they are hard to access and their lucrative contents require much processing to extract. Mining the nodules, however, remains one more catastrophe to add to the other catastrophes visited on the ocean. What makes it expectable and normal is that (lo and behold) polymetallic nodules have already been gifted to humanity by means of the concept of resources, which frames not only what has hitherto been grasped but anything that is graspable.[6]

A diverse world infinite in beauty, mystery, interdependencies, sheer being, past heritage, and future evolution—a world irreducibly wondrous as a matter-of-fact—is redefined and dissipated into just-being-for-using. Such is the upshot of natural resources, a concept that has its moorings in the human privileging and nonhuman disenfranchising that the Great Chain of Being guaranteed.

The devastation of the ocean, for example, is a direct corollary of conceptualizing wild fish as a resource for harvesting. The anthropocentric worldview reclassifies fish as resources by calling them "fisheries" and "fish stock," concepts that get the living beings themselves out of the way and render them instead as protoplasmic substance for mass consumption.[7] As author Ted Danson points out, the fishing industry tracks down its resource with the exact same mindset that the oil industry tracks down its resource—"pushing on in the same fashion, driving their boats farther out into . . . the deepest waters on Earth,"[8] continuously moving to new grounds and eventually other species, when "fisheries" and "fish stocks" run dry. Linguistic categories such as fisheries, fish stocks, and marine resources do not refer to real ontological entities. There are fish and other living creatures in the ocean—there was a huge abundance of them before industrial fishing decimated them. Human-possessive monikers pretend to refer to something in the world, while slipping in a confabulated representation that enables, as merely conventional, colossal extermination acts.

By the time a person learns the words "fisheries" and "fish stock," she is already fully versed in the human-supremacist worldview; she receives an additional confirmation. "Fisheries" and "fish stock" embed possessiveness—the thing already grasped—without the grasping, as such, ever crossing the threshold into self-conscious awareness. The violence this promises to fish and other marine life has been delivered in spades and continues briskly, yet is concealed from awareness because these resource-laden terms inform that fish are for mass consumption and that harvesting them for that purpose is in line with nature's order. In tune with this construction rides the convenient fiction that fish are mindless beings (a kind of animated protein), who are not subjects of their lives, do not experience pleasure or suffering, and cannot feel pain. This is how linguistic constructs and beliefs participate in upholding human supremacy and concealing its mass violence against sentient beings—right in the open.

Similar reasoning applies to "fisheries collapse," an event that has befallen species after species. Such collapses are rarely comprehended as a problem for the fish or for the biosphere or for the living ecologies the fish inhabit. Somehow, the disaster of "fisheries collapse," visited by modern humans most especially, does not befall the beings themselves—the herring, the groupers, the alewife, the tuna, the marlin, the swordfish, the totoaba, the sea bass, the haddock, the cod, the abalone, and so forth—the disaster is regarded as befalling people. Herein lies the twisted genius of human supremacy, which accomplishes much of its work backstage, by means of "language surrendering itself to our mere willing and trafficking as an instrument of domination over beings."[9]

Western classical thought, followed by Judeo-Christian theology, championed the idea that everything has come into existence, or been specially created, for human beings. (The denigration of nonhuman nature from antiquity onward was not without dissenting voices, but they were the minority.) "If nature makes nothing in vain," Aristotle mused, "the inference must be that she has made all animals for the sake of man."[10] "For whose sake will anyone say that the world was created?" wondered Roman author Cicero. "Presumably," he pontificated, "for those animate creatures which use reason: that is for gods and men." Such classical anthropocentric thought influenced "the entire subsequent history of European thinking."[11] Taking his cue from classical thinkers like Aristotle, Cicero, and others, Saint Thomas Aquinas opined: "The imperfect are for the use of the perfect: plants make use of earth for their nourishment, animals make use of plants, and man makes use of both plants and animals. It is in keeping with the order of nature that man should be master. . . . God has subjected all things to man's power."[12] The borders of this hierarchical order, carefully delineated by Christian doctrine with man at the apex of Creation, were crisp and well patrolled. In case anyone might miss the memo, the ancient chimeric deity Pan was duly refashioned into the despicable Devil.

During the Middle Ages, under the spell of Christian dominion-of-man dogma, a rigid anthropocentrism reigned. By the early modern period, it was simply and uncontestably conventional to regard all non-humans and places "as subordinate to [man's] wishes."[13] Thus classical and Christian conceptions forged the foundation for the modern secular concept of resources to emerge and go viral. In the modern era, and in our time especially, otherworldly rationales for the notion that nature was made for people no longer have consensual traction. Yet that identical notion has been upheld in the idea of natural resources—and its many spin-offs like fisheries, lumber, game, livestock, and freshwater, as well as ecosystem services, working forests, and natural capital—which conceptually codify "subjecting all things to man's power" and perpetuate humanity's assumed prerogative to the nonhuman realm (a prerogative previously grounded in metaphysical or theological rationales).

In modernity, human entitlement over the natural world was not proclaimed as God-given; it was, in a manner of speaking, seized. The seizing has been concealed inside pseudo-objective, widely shared linguistic renderings and corresponding technics, which directly substitute for God's decree regarding human privileges vis-à-vis the nonhuman world. Indeed, when the philosophical architects of modernity averred that nonhuman nature is mere mechanism and materiality, and that man is sole rational Subject (René Descartes) and destined master of nature (Francis Bacon), Anthropos was shrewdly set up to usurp God's position. Early modern philosophers crystallized that in the modern era "what is, in its entirety, is now taken in such a way that it first *is* in being and *only is* in being to the extent that it is set up by man, who represents and sets forth."[14] From an ecological perspective, what this statement points to is the post-Enlightenment usurpation of the biosphere—a usurpation that was and continues to be simultaneously representational and physical.

In remaking himself, in secular guise, as sovereign subject who "represents and sets forth," modern man carried forward the legacy of nature domination: no longer as a bequest granted him by a supreme being in whose image he was made, but as an authority he seized and claimed to be intrinsic in his godlike being. Not only did otherworldly rationales become superfluous, eventually so did pompous ideologies in favor of establishing "the empire of mankind" (Bacon). Even as over time human empire was indeed established, it became unobtrusively locked down in sundry representations paired tightly with corresponding technologies. Empire became inscribed in language when nature was named resource, and it was hardened into material cultures via technological means and techniques targeting those resources.

Representing and setting forth—language and technology—have taken on the function of scriptural benedictions and ideological pomp while dropping all overt legitimation requirements. The alliances of fisheries and the trawl, freshwater and the dam, livestock and the assembly line, pests and chemicals, timber and the saw and sawmill—these alliances, as a matter-of-fact, put beings and places at man's disposal, and their amenity to being at man's disposal attests, self-evidently, to the sovereignty of man.

As the natural world's ontology is remade, the remaking as such gets reified within the collective's shared representations and experience, while Earth's original ontology becomes forgotten. And "like the earlier view of human beings created in the image of a God or gods," philosopher Gary Steiner notes, "the modern secularized view still conceives of humanity as . . . godlike. . . . The traditional assumption of human divinity and the resulting sense of superiority over [nonhumans] remain unshaken."[15] Those who are thus fond of calling humanity "the God species" are hubristic only in proclaiming as much aloud: for this idea constitutes the founding belief and ongoing mode of operation of modern culture. Openly calling humanity the God species echoes and validates modernity's replacement of God with man, which has worked as the means to continue dominating nature with impunity.[16]

To say that the prerogative to use nature as humans will was seized in modernity is not entirely accurate: that prerogative can be compared to a baton unbrokenly handed down from the remote past. The historical debt disappeared by being buried in language. When the concept of natural resources is evoked in spoken or written word, it appears as though the world is being described when in actuality the hierarchical narrative of the Great Chain is being echoed and re-etched into the human mind. Reconfiguring the world qua natural resources makes human supremacy actionable, thereby maintaining that worldview as de facto reality, while elevating the possessive agenda inherent in the concept of resources into a sound guiding principle of action.

The idea of resources sets up the nonhuman world as an instrumental field stripped of its interiority, sentience, luminosity, and self-being. This setup is axiomatic to civilization as we know it. As philosopher Richie Nimmo explains, "the distinction between humans and nonhumans is a condition of existence of modernity as a form of order, and indispensable to its continued authority."[17] Civilization's regime of human-nonhuman apartheid remains invisible to mainstream culture, because of the layers of history that have entrenched it, thereby making it commonplace. To paraphrase Nimmo, civilization is built upon the subjugation of the more-than-human world. Subjugation is not an adventitious feature that might be corrected or reformed by, for example, more environmental laws, roomier cages, better management, or green energy. Because nature's domination is the sine qua non of civilization, we must remake civilization in its totality.

Techno-Managerialism to the Rescue

As long as no adverse repercussions accrued to halt humanity's march, the consequences of expansionism were either unperceived or regarded as unproblematic. Yangtze dolphins, western black rhinos, Javan tigers, passenger pigeons, Carolina parakeets, aurochs, gastric-brooding frogs, golden toads, thylacines, dodoes, great auks, Pinta Island tortoises, elephant birds, quagga zebras, Steller's sea cows, Caribbean monk seals, Stephens Island wrens, sea minks, Falkland Island wolves, ivory-billed woodpeckers, bubal hartebeests, laughing owls, Pyrenean ibexes, huia birds, and Tecopa pupfish, among so many other known and mostly unknown beings, have been terminated from the face of the Earth without such losses being perceived as a problem. Populations of wolves, big cats, bears, large herbivores, and countless others have dropped precipitously, yet these declines have not been deemed problematic either. The numbers of fish, sea turtles, and marine mammals have taken a nosedive, while forests have fallen, desertified regions expanded, topsoil evanesced, and rivers and lakes impoverished or rendered lifeless. Indigenous peoples, ways, and languages also became, and are going, extinct. For a long time, none of these consequences were regarded as noteworthy. The price of domination has been, borrowing eco-theologian Thomas Berry's metaphor, collective autism.[18]

Yet today a new development is under way: a growing recognition of the need to contain the adverse side effects of human expansionism. This recognition has arisen because the disregard of limitations to economic growth, population increase, industrial food production, energy use, and sprawl of industrial infrastructures has started to backfire on humanity. The advance of the civilized enterprise—hitherto unproblematic in only having consequences for dismissible others—is ramifying in ways that are jeopardizing people and perhaps civilization as a whole: rapid climate change, freshwater shortages, unprecedented forms of pollution, and topsoil depletion are some high-priority issues for human welfare. Portents of eco-disaster on humanity's horizon are a signal feature of our time.

The prevalent response to ecocatastrophes, however, has not been to confront the worldview driving the blotting out of innumerable beings in blitzkrieg campaigns or by a thousand cuts.[19] Virtually no mainstream politician, media outlet, or organization has contested the historical legacy of humanity's declaring itself supreme, displacing nonhumans and wilderness, and reinventing Earth as civilization's stage set. Rather, the typical response to ecological problems is a riff on the anthropocentric narrative. Foremost focus centers on how humanity's impact might affect the fate of humanity itself. "Human driven changes to the environment," analysts worry, "are raising concerns about the future of Earth's environment and its ability to provide services required to maintain viable human civilizations."[20] Such concerns are often followed up with exhortations about the unique strengths of the human—our intelligence, resourcefulness, and innovative esprit—through which we might resolve difficulties on our march.[21]

Implicitly confirming instead of challenging the human supremacist worldview, the dominant framework for dealing with environmental fallout is piecemeal and techno-managerial.[22] The piecemeal approach treats each problem in isolated fashion, instead of viewing mounting problems collectively as symptoms of unrestrained expansionism. The pragmatic pitch, or enthusiastic push, for techno-managerial solutions rests on the received conviction of human exceptionalism. Indeed, today's most common spin on the Differential Imperative—the axiomatic status of human distinction—is perdurable faith in the technological and managerial prowess of our species to carry the human enterprise onward and forward.

Techno-scientific reports and popular media showcase a host of ostensible solutions to ecological conundrums framed in a compartmentalized and techno-managerial register. For example, scientific and policy reports aver that shortages of freshwater for agriculture, industry, or urban uses might be tackled via desalination, or by other mega-engineering projects like rerouting entire rivers. Diminishing fossil-fuel reserves will be countered (and are being countered) by deploying high-impact technologies that flush hard-to-access deposits from deep-sea and land sediments, mountaintops, boreal forests, and other landscapes. Down the road, researchers hope to repurpose algae, switchgrass, or other biomass into fuel and perhaps food. Should climate disruption become ominous, humanity might rely on geo-engineering wizardry—a newfangled "Star Wars Initiative" against a global-warming offensive some tipping point or other might launch.[23] Adequate food production will be secured by efficient management of land, water, fertilizer, and pesticide inputs, as well as by corporate re-arrangements of crop and animal DNA, so the former can grow on arid, degraded, or flooded lands and the latter made to balloon into bigger "protein" portions. As wild fish are depleted, industrial aquaculture operations can be (and are being) escalated; and since the ocean fish needed to feed farmed fish are running out, no doubt someone is working on how to fatten confined fish on soymeal, corn, or chicken.

In framing environmental challenges as piecemeal technical and managerial challenges, the big picture within which problems are swelling disappears from the terrain of questioning. Since humanity's dominance is not on the table for scrutiny, but on the contrary remains the foundation for moving forward, the mainstream approach is to frame problems within the established order and apply single-file interventions. Unsurprisingly, "innovation" is the buzzword of the day. Large-scale research efforts and financial investments are poured into technological inventions, techno-fixes, and efficiency gains.

The techno-managerial paradigm is pervasive and its brainchildren are copious and multifarious: ranging from plans to redirect rivers like the Ganges and Brahmaputra in order to irrigate agro-industrial landscapes, to parody-sounding techno-fixes such as "Robochop," a proposed remote-operated fleet of robots for mashing up jellyfish blooms (caused by humanity's massive disruption of ocean life).[24] To give a third example, in August 2015, in the midst of a California drought, Los Angeles officials had the city's water reservoir blanketed with ninety-six million black, four-inch plastic "shade balls" to prevent evaporation. "This is a blend of how engineering really meets common sense," one of the decision makers opined. "We saved a lot of money, we did all the right things."[25] Diverting India's holy rivers, engineering jellyfish exterminators, and covering city drinking water with plastic balls: These are examples of initiatives through which the mainstream's techno-managerial approach endeavors to deal with the Pandora's box of plagues that human expansionism has unleashed. A planet inundated with gigantic engineering works such as mega-dams, desalination plants, and nuclear power facilities, but also gimmicky techno-fixes and Band-Aid solutions that sound like *Onion* editorials—this is the new world order coming online.

Techno-managerialism is itself the legacy of the secularly established "empire of mankind" that the early modern period launched. The contemporary technological and managerial framework epitomizes this psychohistorical inheritance. Over the course of time, the more the nonhuman domain has come under man's disposal, the more confidence have humans come to feel in their power to calculate, engineer, plan, mold, and solve. At a microlevel, confidence in the framework manifests in the way sundry problem sets and challenges are taken up, serially and diligently. At a macrolevel, the techno-managerial framework's confidence manifests in the aura of haughty invincibility that envelops it—undaunted, as it apparently remains, before the dire specters of rising seas, scorching heat waves, shrinking fossil-fuel reserves, species extinctions, long-lasting pollutants, deadly viruses, and the like. Problem-specific and endlessly innovating techno-managerialism broadcasts the missive that there exists no symptom it cannot solve—a missive most especially loud and clear in its refusal to tackle, or even look at, the root cause of all banes.

The techno-managerial framework sustains human colonialism, while angling to contain damage and effect some cleanup. Whether or not it could succeed in solving or muddling through the immense quandaries humanity faces, the outcome is bleak: If techno-managerial approaches fail in the mission to address freshwater shortages, "feed the world," or mitigate dangerous climate change (to mention three looming issues), the result will be unthinkable suffering. If techno-managerial approaches succeed in containing dire side effects for humanity, while entrenching human planetary domination, the cosmic gem of Earth will be turned into a downgraded resource base to serve human users.

Just as there is an underlying driver of the problems we face (a human-supremacist worldview that has underwritten the biosphere's colonization), so there is an underlying pattern to the mainstream solution framework (a diligent avoidance of calling out that worldview). The prevalent techno-managerial framework for civilization-endangering problems zooms the focus on, and even kindles admiration for, the cleverness of Anthropos in the face of adversity. By the same token, this approach safeguards the belief that Earth is the Planet of the Humans: by never allowing that belief to be unveiled for conscious scrutiny and rejection; and by rehashing already-acclaimed superior human qualities as those to be marshaled for moving forward. By silently summoning the supremacist worldview to the rescue, the dominant framework bypasses unmasking that worldview as the very source of all the problems that managers, technocrats, innovators, securitization experts, and specialists of all sorts will try to solve.

Indeed, techno-managerialism seeks to make the anthropocentric regime sustainable and thereby turn Earth's occupation into a success story of civilized humanity. To wit: that Earth belongs to humanity;

that the planet consists in resources and services for human advancement; and that human beings are exceptional by comparison to all other species. The spell of this belief system seems so soporifically potent that the civilized establishment will sooner assume charge of the planetary climate system, or swallow the pill of mass extinction, than relinquish that belief system.

Ironically, the steadfast loyalty to human supremacy is neither conscious decision nor deliberate strategy. Just as the engine of the anthropocentric worldview has swallowed up the world in its compulsive expansionism—regurgitating impoverished landscapes and ecological amnesia in its place—so has it swallowed up humanity.[26] Humans shaped the world in accordance with a worldview that has conditioned their own perception and constricted their own understanding. Otherwise stated, people are as much a casualty of that worldview as the biosphere. For while human beings have indeed wielded jaw-dropping technological works to unlock nature's wealth, they have never been in control of the ontology that nature's domination produces.[27] Plundering the biosphere has transformed much of it—and ultimately will transform all of it—into a realm that reflects the domination project back to the human mind. People experience this biologically degraded world as the real world, and are inexorably propelled to continue acting according to its dictates. Thus has the creed of superiority and entitlement ensnared humans in a feedback loop: it incited them to conquer and humanize the natural world, and the conquered and humanized world continually reinforced the supremacist credo. This vicious cycle is the snare that humanity cannot seem to escape and the box many people cannot even think outside of.

Humanity appears locked into an expansionist way of life. Even as the natural world is converted and folded into a human-user domain, so are people confined inside an increasingly narrower ontological groove through the erasure of ecological dimensions of the living world through which they would be set in motion, cultivated, in an alternative way. Superior and entitled humans end up living banally on the Earth, and the kind of Earth that might still enjoin them to "dwell poetically"[28] is vanishing. "If nothing succeeds like success," philosopher Hans Jonas astutely remarked, "nothing also entraps like success." He proceeded to clarify this statement by observing that "the expansion of [man's] power is accompanied by a contraction of his self-conception and being."[29]

Herein lies the boomerang of anthropocentrism: while it wagered on reaping power by elevating humanity as the crown of Creation and turning the world into its oyster, it has so conditioned the human that the possibility of a way of living rife in multispecies flourishing and vibrant with diversity and abundance of beings has almost been rendered unthinkable. It is thus not so much that people insist on upholding anthropocentrism; rather the long history of anthropocentrism seems to have taken away humanity's freedom to renounce it. The worldview speaks, and humans are neither its eloquent champions nor, any longer if ever, its privileged beneficiaries; they are its instrument.

Human Supremacy as Sociohistorical Construct

Humanity's domineering presence has long been catapulting the biosphere toward our present-day predicament. This Earth-colonizing venture, however, has not been a consequence of who we are: it has been a sociohistorical outcome, albeit one that humanity became increasingly unable to escape, since over time people became trapped into revamping the world in accordance with the belief system of superiority and entitlement. Human beings have treated the world as though the conception of nature as "standing reserve," and the sedimentation of a human-nonhuman hierarchy, were ontologically sound and ended up creating a man-made ontology that appears to display the legitimacy of human posturing, and encourages people to continue deploying techno-managerial systems toward completing the transformation of Earth into a resource base.

By now, humanity no longer requires religious or philosophical justifications for dominion. Having rehearsed numerous iterations of the overlord story for millennia, and having acquired the physical power of overhaul and surveillance of the planet, the human enterprise has turned itself into empire. Through accrued and snowballing momentum, geographical expansionism and its attendant Differential Imperative ideas have sponsored the anthropological colonization of the living world. Human supremacy hardly needs a supporting narrative, as it has been congealed and ossified into language, institutions, and conventional activities. And so, in the spot-on words of the anonymous authors of *The Coming Insurrection*, "when all is said and done, it's with an entire anthropology that we are at war. With *the very idea of man*."[30] Planetary takeover has been orchestrated by a conditioned humanity that not only invented empire after empire, but ultimately has succeeded in defining our species-being—the very idea of man—as empire.

Human supremacy, as an enduring ideational-pragmatic historic construct, is the underlying engine of impact. This understanding has remained relatively unavailable, because of an abiding disconnect between histories of cultural ideas about nature (on one hand) and histories of human activities upon nature (on the other). In our libraries there exists extensive scholarship about the history of anthropocentric ideas (from antiquity to the present) and about the history of regional and world ecological degradation (from Sumerian deforestation to the thawing poles). Rarely are the two storied in unison in any single work.[31] The riddle of history solved lies in connecting the dots, for culturally dominant beliefs are never separate from reigning institutional arrangements and actions. The historical unfolding of human-supremacist ideas coupled with subjugating actions have been the two strands of history's loom that, through their unbroken shuttling back and forth since civilization's birth, have led to the present moment of a human-occupied Earth. The weaving pattern continues tenaciously to hold.

Vainglorious conceptions of the human did not just thrive in ivory towers of academes, learned circles of scholasticism and theology, or lofty enclaves of scholarship and power elites. Such ideas trickled down to the masses, and were indeed actively disseminated from church pulpits and through centers of learning into the social body: crafting culture, organizing perception, fashioning language, motivating action, sculpting physical reality, and ultimately forging the human identity of (what philosopher Immanuel Kant applaudingly called) "lord of nature"[32] who, still, mechanically strives to carry the bankrupt legacy of human-centeredness forward. As authoritative and omnipresent as it appears, the lord-of-nature identity is not rooted in human nature in any essential or final way. Just as it has been historically constituted, so it can and must be historically undone.

Unmasking the worldview of human supremacy as a conditioning regime gives space to the imagination to see ourselves anew as children and inhabitants of a biosphere abounding in the closest thing to pure magic that we know. Life! Thought freed from the strictures of human distinction and special prerogative turns toward another worldview to live by: one in which human beings—whether ultimately they choose to preserve a diversity of cultures or to embrace a melting pot of diverse individualities made from collages of diverse cultures—thrive within the biosphere's living plenum, with Earth freed to create exquisite compositions of being.

PART TWO

Discursive Knots

FOUR

Is the Human Impact Natural?

Civilization was purchased by the betrayal of Nature. E. O. WILSON

In recent decades, a raft of findings has generated a comprehensive understanding of the unraveling and revamping of the biosphere under way. Climate change, with its already visible and experienced effects, has played a pivotal role in stimulating research into global land-use patterns, freshwater extraction, deforestation trends, species extinctions, ecosystem losses, disturbance of biogeochemical cycles, and other impacts.[1] Some scientists and analysts contend that the massive consequences of our dominance warrant adopting a new name for the epoch we live in. The name "Anthropocene," they argue, is needed to convey that Earth has exited the envelope of natural conditions and variability of the Holocene (the epoch following the Pleistocene at the end of the last glaciation about twelve thousand years ago), and entered into a human-instigated biogeological age.[2]

Humanity's impact poses an imminent threat not only to Earth's biota, ecologies, and longstanding topographies, but to a safe future for people and future generations as well. As scientist Anthony Barnosky and his coauthors put it, humanity is forcing a "planetary-scale critical transition . . . with the potential to transform Earth rapidly and irreversibly into a state unknown in human experience."[3] Human actions are perturbing the world so swiftly and drastically that scientists advise to "expect the unexpected," given that "the plausibility of a future planetary state shift seems high."[4] Despite such warnings human expansionism continues, and potentially colossal threats remain largely unheeded. Ongoing population increase, growing economies, and infrastructural sprawl—the large-scale "trends of more"—tend to be treated as the unchangeable variables around which technological fixes, efficiency tunings, and sundry adaptations must be applied.

While clear discernment of human dominance and its menacing effects seems a hopeful development, emergent environmental discourses (especially those orbiting around the Anthropocene concept) have failed to inspire and mobilize the energy to relinquish that dominance. Instead, there is a prominent tendency within much environmental thought to accept the domination of nature as given. As discussed, techno-managerial approaches predominate, while advocacy to halt the trends of more—by means of stabilizing and substantially lowering the human population, shrinking the global economy, and pulling back from huge swaths of the biosphere—remains scant in the mainstream.[5] The question that naturally arises is why, despite looming dangers, human dominance and continued expansionism are not on the table for questioning.

A most general response to that question is that the worldview of human supremacy is deeply ingrained and largely invisible in the social body. Relatedly, the constitution of the world as a supermarket of natural resources serves powerful political and corporate interests that are, more often than not, impervious to change and protest. Such historical and socioeconomic factors are unsurprising aids of the status quo. More surprisingly, the supremacist worldview is also underquestioned within the environmental arena, which has as its mission dissecting the human-biosphere relationship, identifying what has gone awry, and offering pathways to a better future. The question of why environmental thought has desisted from confronting anthropocentrism and failed to agitate for abolishing nature's domination deserves scrutiny. This chapter along with the two that follow investigate certain "discursive knots" blocking the way to such questioning.

The metaphor of discursive knots is borrowed from Buckminster Fuller's description of a knot as an "interfering pattern," one that becomes harder and harder to undo the more knots are piled and tightened on one another.[6] Similarly, the discursive knots I critically examine are patterns of thinking about the current situation that impede the flow of thought and action in an alternative way forward—of contracting the human enterprise rather than adjusting to its expansionist crusade. The more interfering thought patterns are repeated, the more entrenched they become, and the more they appear to be "truths of the matter."

A widespread belief that humanity's impact is a direct upshot of "human nature" is such a discursive knot—one of Gordian proportions given how well founded this belief appears to be and how widespread it is. It is a belief that, intentionally or inadvertently, lends support to the established order of human dominance. For if the latter is a corollary of our peculiar-exceptional biological nature, then the inference encouraged is that we must either embrace or resign ourselves to some form of human planetary rule (and that the best we can do is "green" that rule). The belief that planetary dominance supervenes from "who we are" seriously disincentivizes humanity from relinquishing its reign.

The conviction that the human domination of nature is an inexorable consequence of our inborn distinction naturalizes humanity's impact. As founded as this belief may appear to be, it does not hold up to sustained scrutiny but is disputable on a number of fronts. First, human action is always sociohistorically situated, so its expressions in the world, in the words of critical theorist Judith Butler, "must be understood as the taking up and rendering specific of a set of historical possibilities."[7] Naturalizing the human impact systematically sidelines the fact that human action is always entwined with shared ideas and attitudes, which, having been culturally instilled into the social body, furnish the prime directives of conventional actions. (The idea itself that humanity's ascendancy flows from inborn human traits is also culturally favored—a point to which I will return.) Naturalizing the human juggernaut also disregards enormous cultural and individual variabilities in how people relate, and have in the past related, to the nonhuman realm. Additionally, naturalizing humanity's impact cannot explain why the ecological crisis raises ethical and existential questioning at all—let alone questioning so adamant that it will never be silenced. As a last point, the human-impact-is-natural view cannot account for the profound grief that countless people worldwide are experiencing in response to the onslaught on the natural world. These dimensions of the human condition—of social conditioning, cultural and individual variability, deep questioning, and grieving—implacably complicate any facile notion that planetary dominion supervenes from our species' makeup.

Yet the problem with the belief in the naturalness of human impact goes even beyond the fact that it is highly dubious. The deepest problem with this belief is that it encourages the conflation of the conditioned identity of nature colonizer with human nature and promotes the reification of humanity as planetary owner, thereby buffering the human-supremacist worldview while making it "magically" disappear. To inquire into the anthropocentric mode of operation as sociohistorically constituted is to disclose that it is *one* constructed meaning of being human. This disclosure opens a horizon within which we become free, or at least freer, to work toward recreating who we are and our way of life by charting a new historical course. But humanity courts the danger of losing the freedom to remake itself as an integral member of the biosphere, if it buys into the pitch that its identity as overlord is natural.

In terms of the question of who we are as a species (if that question even has a universal answer), the reality is that this is precisely the question that, amidst reckoning with a shattered world, confronts us. Any interpretation of the human that already presumes to know "fundamentally what man is . . . can never ask who he may be."[8] All fixed ideas of what human nature consists in will systematically block inquiry into who the human may become. It is within the horizons of our imaginative-pragmatic capacity that we may become a people who abdicate human empire instead of struggling to ensconce it and just clean up its self-endangering corollaries. A currently in-vogue notion that our species' distinguishing powers endow us with godlike stature sabotages the potential realization that humanity does not have to reign in the biosphere to achieve greatness. Indeed, reinforcing the belief in human distinction impairs the capacity to recognize that the path to human greatness lies exactly in the opposite direction: in allowing the biosphere's magnificence to be what it is, while embarking on the quest of who we must become in order to inhabit and be enriched by that magnificence.

Many human-driven impacts are clearly irreversible, including innumerable and mostly unknown species extinctions, a climate-disruption episode that might still be mitigated but not entirely reversed, and global contamination by long-lasting pollutants like toxic chemicals and plastics. Yet from the reality of certain huge or irreparable blows, it does not follow that we must accept human dominance as fait accompli. Even so, this is the prevailing message of much environmental thinking. On the other hand, promoting the tack of scaling down and pulling back the human enterprise is the domain of "minority reports," which often get dismissed as radical or nonpragmatic.

I critically dissect three discourses in contemporary environmental thought that naturalize the human impact and thereby subtly promote ensconcing the domination of nature. The chosen discourses are important on a couple fronts. They focus on the long haul and deep history of the human shaping of the biosphere, and thereby appear to make an indisputable case that the ecological culprit is *Homo sapiens*. Additionally, the examined discourses are prominent within and outside the environmental space, and thus influential in shaping views in academic, public, and policy-making circles. To the extent that such discourses encourage (or help create) a consensus that our predicament is "natural," the motivation for fundamental change becomes vitiated.

First, I examine the model of "land-use change" or "land-use transition" to describe the postagricultural unfolding of human expansionism across the globe. This model, I argue, represents the historical conversion from wilderness to human-dominated landscapes as a natural process instead of one driven by specific sociocultural forces and formations. Next, I consider a pervasive inclination in the Anthropocene literature to depict human ascendancy in the biosphere as a natural event—indeed, as the current chapter of Earth's natural history. Last, I focus on a specific (and typical) portrayal of "Pleistocene overkill," which contributes to naturalizing humanity's onslaught by representing the extinction crisis as a single continuous event from the prehistory of our species all the way to the present moment.

Representing Earth's Takeover as Stages in Land-Use Transition

Earth's fate in the Holocene—the period within which agriculture and civilizations emerged—is often captured through a model called land-use change or transition. This describes the spread of the agricultural mode of production across the globe during the last roughly eight millennia. A graph published in a 2005 *Science* article captures, at a startling glance, the terrestrial changes effected by humans in this time span (figure 1).[9] As agricultural societies spread to new lands, they effected "frontier clearings" such as deforestation and other natural vegetation conversion. Wildlands were repurposed for crop cultivation and animal grazing. The first three stages represented in the graph (presettlement, frontier, and subsistence) were rehearsed repeatedly in different places, ultimately scaling up to the final two stages (intensifying and intensive), which sum up the state of the globe today.

The image relays a geographically stark depiction of how the terrestrial biosphere has "transitioned" from being governed by wilderness to being dominated by agriculture. (More recently the ocean has

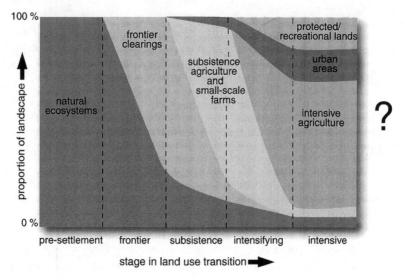

Figure 1   Stages in land-use transition. (Source: J. Foley et al., "Global Consequences of Land Use," *Science* 309 [2005]: 570–74.)

undergone an analogous shift from life abundant to fished out and polluted.) While the model is descriptively valid, by eliding its sociocultural contexts it tends to imply, or encourage the perception, that land-use transition has been a natural unfolding of *Homo sapiens'* sojourn on Earth. By containing both messages—one factual, the other pseudo-factual—such representations of land-use change simultaneously disclose and conceal the character of the human postagricultural encounter with the land.

Even as land-use transition accounts expand our knowledge of environmental histories by empirically showing, as a more recent paper puts it, that "land use has been extensive and sustained for millennia,"[10] such accounts typically abstract from the sociocultural circumstances and formations that drove the changes. Thus while the encounter with environments is historicized in ways that are illuminating, the social dimension of that encounter is misleadingly dehistoricized. As philosopher Jacques Derrida described this type of discourse, "the history of the concept of man is never examined. Everything occurs as if the sign 'man' has no origin, no historical, cultural, or linguistic limit."[11] Along the lines of this insight, when narrated environmental histories are evacuated of the social conditions that forged those histories, it appears as if the forging were done by "man" qua natural entity, devoid of historical, cultural, and linguistic (i.e., ideational) ground.

Yet the human beings who produced those environmental histories, breaching frontier after frontier, invading the wilderness and any (indigenous) peoples in the way, were conditioned by and acted within specific contexts of social structure and organization. These social contexts codeveloped with the agricultural mode of production and reproduction. Those were the societies within which, early on, hierarchy, militarism, stratification, and the conditions (and eventually conceptions) of "wealth" and "poverty" came into existence. As these societies burgeoned over time—both in population numbers and organizational-technical complexity—they morphed into city-states, empires, and nation-states. These social formations consolidated, and for as long as they could persist required, slavery, peasant serfdom, armed forces, state repression (by force or taxation), looting expeditions, colonialism, and large-scale warfare. Conquest and (depending on historical circumstance) genocide were among the foremost, and (sooner or later) necessary, means of expanding animal and crop agriculture into new terrains.

In hindsight, how the dynamic between complex agricultural societies and the natural world unfolded may have been overdetermined, so to speak, what with the reinforcing interplay of eventual land degradation, inexorable human and livestock population increases, and insatiable elite ambitions for wealth and power. Yet even with a level of inevitability built into how the history of agricultural humans played out, that historical process was not "natural." Indeed, it has been a process entirely premised on the sociocultural programming of a specific human identity—one separate from nature, superior, and authorized to take by fiat. Those societies birthed that identity, strengthening and marshaling it century after century via a portmanteau of armies and weaponry, excursions of plunder and occupation, and sundry spectacles of murderous gore for the masses.[12]

Periods of calm and pockets of well-being interleaved the modus operandi of conquest, especially when the land yielded the proverbial milk and honey, olives and fruits, bread and wine. But from our vantage point, the agricultural juggernaut originating millennia ago in the Near East—a juggernaut that happened to consist in a portable ensemble of livestock, crops, and (slowly changing) technologies—spread by means of takeover throughout the Mediterranean basin, Eurasia, all of Europe, the New World, and after the Industrial Revolution (and especially after 1950) to the whole world.[13] Agriculture emerged contemporaneously (and independently) in a handful of locations around the world at the dawn of the Holocene. However, the agricultural system that came out of the Near East was the one that developed the most burgeoning livestock component, which contributed immensely to its distinctive aggressive expansionism.[14] The West's "ecological imperialism"[15] was spearheaded by the exercise of extreme force against wild nature and native people who happened to stand in the way. Yet extreme force does not explain much if there is a failure to appreciate that those who exerted such force were inculcated into a belief system that the natural world and its inhabitants (wild, domesticated, and indigenous) are for use or extermination.

Returning to the "stage in land-use transition" image, we see how it speaks a thousand words while also concealing much. It speaks a thousand words: Agriculture has virtually occupied the biosphere. It conceals much: Even as the human wave that etched that history on the land (and ocean) is dehistoricized, so the relation between humanity and the natural world is depoliticized. It is depoliticized in the sense that land takeover is represented as an induced alteration rather than the establishment of a totalitarian regime by a certain type of human.

The depoliticized relation can be elucidated through clarifying what the model is *not* portraying. It is not portraying human action motivated by a desire to share the world and cocreate it with other beings and processes. It is not depicting people devoted to the pursuit of mutual thriving among all earthlings. It is not showing a process by which humans endeavor to make a home within the sacred community of life. It is not portraying a relationship between human beings and wild nature, but at a fundamental level the lack of one—or more precisely, the pointed avoidance of one. While the image offers a clinical description of what has occurred, interred within it is that "land-use transitions" have invariably been achieved by means of extinguishing, displacing, persecuting, killing, and exploiting the natural world and its residents. As social historian Jason Moore characterizes this type of exposition, the land-use transition model is "descriptively powerful perhaps—but analytically anemic."[16]

The transition from biosphere-wide wilderness to biosphere-wide agriculture is testimony to the increasing loss of freedom of the planet's community of life. Destroying freedom is how civilization and its mega-agricultural model—and now its epiphany of industrial agriculture, industrial fishing, and industrial aquaculture—have proceeded. In contemplating the land-use transition image, we are not so much looking at a "photograph" of what has transpired on Earth as we are seeing how human-supremacist imperialism has been operationalized.

The mode of operation undergirding the model has not been ordained as a natural historical process. It is a mode that has been mediated by a prevailing worldview, a way of proceeding shaped by powerful ideas about, and associated economic and technical cultures aimed at, nature and especially wilderness. For example, for Europeans of the Holy Roman Empire through the early modern period, deforestation stood for the triumph of civilization and was carried out according to a "long tradition that to strike down trees was to strike a blow for progress."[17] In the Middle Ages (especially after the invention of the plow), powerful rulers (lords, princes, and the like), anticipating future profits, financed the expansion of agriculture into still-wild forested plains, moors, and marshes of Europe, which prior to their occupation and domestication were labeled "deserts."[18] Later, Europeans brought their traditions of nature plunder and wilderness loathing to the New World. "The woods," declared settler Cotton Mather, are "to be cleared of . . . pernicious creatures to make room for a better growth."[19] Virginian colonists, having transported their supremacist culture across the Atlantic, offered the reward of one cow for every eight wolves killed.[20]

The broader point being that cultural ideas and attitudes have never been empty musings floating above the real world. The chronicles of Western thought are replete with exhortations like the above, zealously enacted on the ground. Land-use transition is the end point of the harsh echoes of such ideas; it is as much a model of what has transpired as it is a representation of the enactment of human supremacy. By forgetting that the model has buried within it (as constitutive of what turned it into reality) a certain perspective on the natural world, the model appears as a natural unfolding. Thus the worldview that made (and makes) it reality gets to go scot-free.

The model's entrenchment and supposed inevitability is precisely what we find in mainstream environmental thinking. As long as the mode of operation that generated the model was not harmful for civilization, the model seemed as a sanguine blueprint of action that benefited people. For example, in the second half of the twentieth century, "land-use transition" was unleashed across the planet with the dissemination of industrial agriculture and the launching of its systematic incursion into the tropics.

This way of proceeding—of assimilating the world for the purposes of maximizing food production and aggrandizing the human enterprise, now globally scaled—has sponsored a world conflagration now backfiring on humanity. Out of its cauldron all manner of afflictions have spilled: dead zones and dead rivers, extinct beings and lost ecologies, constant killing of inconvenient animals, global contamination by chemicals and plastics, and an impending mass extinction. Also presently unfolding is the loss of "ecosystem services" for people, with pollinators disappearing, topsoil blowing away, destructive and unpredictable weather, shrinking rivers and lakes, dwindling glaciers, fishery collapses, and eroding coastlines. In other words, a gigantic tragedy for terrestrial and marine nature and now for human beings as well—a tragedy that directly stems from "land-use transition" unfolding toward a global epiphany. But when the land-use transition model is presented as what "naturally" happens when human beings encounter wild nature, then it implicitly follows that dire repercussions caused by the model must be solved within the framework of the model. Naturalizing land-use change disallows fundamental critique: instead, the model appears to call for reforming, and the best means of reform tend to be technological and managerial. Techno-managerial adjustments and innovations try their best to solve adverse consequences in the wake of human domination of nature without *ever* calling domination by name.

Naturalizing the biosphere's takeover is the discursive wand that makes the worldview of anthropocentrism disappear even while openly propagated. It is thus unsurprising that the paper containing this revealing image laments dwindling "freshwater resources" and "forest resources," deploying a language for the natural world as human owned while presenting that language as descriptive. The response to the predicament of dwindling "resources" is to figure out the new technologies or managerial settings that might solve, or at least alleviate, impending crises. Techno-managerialism constitutes the most recent incarnation of anthropocentrism, solving problems generated by the human conquest of land and ocean, and endeavoring to save the phenomenon of biosphere occupation by making it sustainable in the long run.

The relationship between people and nature is always shaped by ideational-actionable constructs, at individual and group levels. The transition from a lush wild world to a food-production dominated biosphere has been orchestrated by the supremacist helix of human superiority and entitlement driving nonstop expansionism across land and seas. When the driving force of environmental change is depoliticized, however, the message of the "natural human" effecting land-use changes is profoundly brightened by the very absence of specific sociocultural color, as it were. Inside this seemingly neutral, "natural" zone even the word "frontier"—a political concept, and indeed the quintessential concept of the conqueror—can be slipped in as descriptor of a transition or change. Other cultures—the ones that for now appear defeated—did not conceive of the places they did not inhabit as "frontiers" calling out for takeover. As Shawnee Chief Tecumseh ("Shooting Star") observed, the idea of frontier belongs to "White People, who are never contented but always encroaching."[21]

Differences in cultural mindsets go a long way in explaining, for example, why Europeans destroyed the Old World's salmon runs in the Middle Ages, and the New World runs more recently,[22] while Native Americans did not destroy the Atlantic or Pacific salmon runs during thousands of years of inhabitation. "The indigenous salmon fishery of British Columbia," write researchers Daniel Vickers and Loren Mc-Clenachan, "was sustained for something in the order of three to five thousand years before European settlement on the Pacific coast, not because Indians lacked the technical capacity to overfish or because they fished for personal consumption only, but because they possessed an economic culture that *emphasized subsistence, diversification, and reciprocity* and because their tribal *culture was not inherently expansionist.*"[23] Odds also are that the Pacific Northwest Indian languages did not have words translating into "salmon fishery."

Native Americans respected the salmon both for their intrinsic salmon lifeways and for the fact that they were, seasonally, food for the people. When the salmon arrived in their natal rivers and streams, the tribes greeted them with established rituals that emphasized the sacredness of relationship with nature's patterns and rhythms, while also demanding restraint and delayed gratification on the part of the people before feasting. "From one end of the vast river system to another," writes ecological restorationist Freeman House, "people were restrained from casual consumption of the fish until certain ceremonial practices were performed in certain locations."[24] While some would blithely label such practices as an "indigenous management of the resource," it is always best, as cultural anthropologists like to enjoin, to describe people's practices in their own words and frameworks of meaning. (I discuss the recent academic fad of calling native people "natural resource managers" in the next chapter.)

In framing the biosphere's predicament, many contemporary proponents of the techno-managerial framework not only ignore the relevancy of sociocultural ideas, but also present techno-managerialism as a hard-nosed pragmatic program. Yet techno-managerialism is itself a mindset: one that diligently avoids challenging the subjugation of nature and covertly glorifies the human capacity to solve all problems. Techno-managerialism puts humanity's dominance on "life support," by bridling the biosphere with sundry mega-technological projects, serial techno-fixes, and endless managerial calculi that aim to clean things up around the edges, perhaps keeping Earth-system variables within quasi-safe boundaries for civilization's onward march.

The Portrayal of Natural Human Ascendancy in Anthropocene Discourse

While the techno-managerial framework is widely embraced in mainstream circles, it has found a particularly spacious home in the framing of our predicament in terms of the arrival of the human-shaped epoch of the Anthropocene. The term "Anthropocene" has proved catchy. While a wide range of environmental writings use the term as shorthand for humanity's alarming impacts, it is also the nucleus of a specific mainstream discourse that actively promotes the term, urges its adoption as the official name of our geological epoch, and is disseminating the perspective of an amorphous, monolithic "humanity" as biogeological force.[25] Historian Christophe Bonneuil identifies four distinct narratives that have clustered around the Anthropocene. My analysis is specifically targeted at what he labels the mainstream "naturalist narrative"[26] that here I call "Anthropocene-think."

Anthropocene-think has become a major hub of prominent environmental trends: identifying our species' nature as the cause of humanity's onslaught on the biosphere; tacitly condoning humanity's overpowering the rest of life by means of naturalizing, and sometimes even eulogizing, human power itself; and promoting technological and management solutions as not only salvific, but expressions of the very essence of our species-being. Anthropocene-think constitutes a major hub of these ideas, because it enthusiastically advances them and because it is an influential environmental-intellectual development of our time.

The way that the Anthropocene literature naturalizes the human hammer are so multitudinous and mutually compounding as to appear to settle the case that human nature is the driver of planetary impact. Yet any seeming persuasiveness on that score is but the smoke and mirrors of discourse, precisely underscoring the point that ideas have extraordinary power. It is important to unmask the discursive pitch to explain the overhaul of the biosphere as stemming from the biosocial essence of our nature. For human nature may not be angelic or a blank slate, but neither is it, by constitution, lord and manager of the manor.

In the popular and scientific Anthropocene literature, today's end point of human dominance in the biosphere is portrayed as demonstrating our species' distinctiveness and special powers. "The Anthropocene," write scientists Paul Crutzen (co-coiner of the term) and Christian Schwägerl, "highlight[s] the immense power of our intellect and our creativity and the opportunities they offer for shaping the future."[27] Similarly, the *Economist*, after noting that the word "Anthropocene" "dramatizes the sheer scale of human activity," concludes with a poignant soliloquy about human intelligence "on a planetary scale" as "something genuinely new and powerful." The same article also remonstrates against an inclination to suspect human narcissism behind such self-assessments, declaring that "the lamentation of vanity can be false vanity." We are encouraged to admit how very distinguished and smart we are, especially as this admission will inspire us to get on with the task of constructing a "smart planet."[28]

A *New York Times* piece on the Anthropocene remarks that "we are the only species to have defined a geological period by our activity,"[29] thus subtly glamorizing human-driven planetary upheaval. This kind of embellishing is typical in Anthropocene-think, both expert and popular. For example, Will Steffen and colleagues support the naturalizing diagnosis of "humankind's brainpower and technological talent in shaping its own future and environment."[30] Statements that causally link nature's destruction with the human brain, intellect, and technological genius are offered in passing, as though self-evident, but sometimes they are garbed in pronouncements that explicitly represent *Homo sapiens* as endowed with characteristics and powers that are nonexistent in other species.[31] Such statements echo the longstanding philosophical, political, and theological trope of portraying our species' ascent into an unprecedented type of entity from out of the mass of the merely living—the age-old Differential Imperative narrative recorked in a new Anthropocene-think bottle. Anthropocene writers like to echo the still-extant—though discredited after the discovery of evolution—motif of "human consciousness as the crowning achievement of evolutionary development."[32]

Nature's domination can, of course, just as easily be interpreted as human beings *not* using their brain-power or extraordinary intellect (including foresight and long-term thinking), constructing and implementing technologies with no regard for other species or future human generations, and acting with neither compassion nor empathy for nonhuman beings. The point being that the latter storyline of a "disconnect" between human nature and human activity in the biosphere is as compelling, if not more so, than Anthropocene pronouncements on our extraordinary endowments as shaping forces in the world. Yet

Anthropocene-think prefers to overwhelm its audience with the register of "wowing" human nature.

The second way that the Anthropocene literature naturalizes the human impact is by representing it on a par with the elemental and geological forces of nature. The wowing of human nature is complemented by a breathless description of its planetary fireworks. *Humans have become a force of Nature reshaping the planet at a geological spatial and temporal scale.* Some variant of this sentence seems obligatory fare in most writing about the Anthropocene. Humanity's onslaught is usually not compared to any specific "force of Nature," but instead the analogy's mystique is sustained by keeping it vague. "Humans are not insignificant observers of the natural world but central to its workings, elemental in their force."[33] The invitation to be mesmerized by the "shock and awe" effect of human power even blinds writers to their odd choice of words: for there is nothing "insignificant" in being observers of this amazing world we've found ourselves in, but virtue and wisdom in such a choice. "Humans rival the great forces of Nature," a *Science* article on the Anthropocene echoes. It quotes geologist Jan Zalasiewicz comparing our time to previous epochal shifts in the geological strata: "I feel quite the same sense of awe when I think about the kinds of large-scale geological changes that we are making to our planet now."[34] The representation of human power using analogies to awesome elemental forces in the universe is the most recent spin on the mythos of human self-glorification—not a mythos that needs rehearsing.

The third way in which Anthropocene-think naturalizes the human impact is by representing the colonization of the biosphere not as colonization but as the current phase of Earth's natural history; that representation is carried forward another step by concretizing Earth's current phase as a biogeological epoch to be named after ourselves. The term "Anthropocene" conceptually crystallizes the diagnosis that "we are now a defining force in the geological process on the surface of the Earth,"[35] making this event continuous with the planetary natural order. The arbitrary, catastrophic exercise of power over a living world that has consistently been deemed inferior and human property, rather than being openly identified as domination, is contoured into a natural history event—a discursive move that profoundly undermines resistance to Earth's colonization.

It is beyond doubt that humanity's onslaught is manifesting at a biogeological level. This has not been an unfurling of Earth's natural history, but the upshot of specific sociohistorical constellations that succeeded in dominating. Anthropocene proponents have coined a variety of neologisms and phrases that naturalize human dominance (discussed shortly), yet prime among them is "Anthropocene" itself, which reconfigures the occupancy of Earth—by a humanity held spellbound by a supremacist worldview—into a biogeological event, while slipping in a boost for this worldview by naming the new epoch after the conqueror. Proponents of the Anthropocene have endeavored to preempt the charge of narcissism inherent in the name by averring that "the Anthropocene will be a warning to the world."[36] This claim seems as founded as the idea that calling our species "Homo sapiens" was an auspicious name choice for encouraging people to cultivate wisdom. On both counts, regardless of stated intentions, human conceit is echoed and reinforced.[37]

The concept "Anthropocene" inscribes civilization's revamping of the biosphere onto the tissue of the Earth. Willy-nilly we are goaded to acquiesce that the planetary upheavals we are living through are but the play of natural history due to the antics of a unique species. It is simply not so. The particular course that history took since the Neolithic has been about the march of a certain kind of civilization that constructed a certain kind of human identity: the supreme-subject identity that is neither our biological heritage nor our universal nature. It is not who we might have been had we created a different kind of civilization; it is not who many indigenous peoples were, before they were either co-opted or stamped out of existence just as surely as the wilderness they co-inhabited; it is not who we all are; and in any case, it is not who we might become in the future. "Species Man does not make history," social theorist Donna Haraway points out.[38] It is not the human species that catapulted the present situation, but a parade of agricultural, militaristic empires—molding human identity and monopolizing the human condition—that have today morphed into a global consumer empire feigning to reign over Earth.

Given its all-around propensity to naturalize the human impact, it is not surprising that the mainstream Anthropocene discourse also favors a neutral language to describe that impact. On virtually every page of Anthropocene-think, we read that human beings "shape," "change," "alter," or "modify" the natural world, concept choices that fly in the face of a more ethical-political language that human societies are "destroying," "overtaking," "colonizing," or "degrading" the planet. The latter ways of wording seem biased by comparison to the neutral language that the Anthropocene discourse flaunts. But using ostensibly neutral language to convey the human impact is profoundly normative in terms of the message it conveys—and thus anything but unbiased.

For the effect of a neutral language is to purge human activity of any ethical dimension in relationship to the natural world, conveying that the lens of "doing right" and "doing wrong" onto others has no application in the human-nonhuman bond. For comparison, consider describing the lives of millions of slaves brought to the Americas during the Middle Passage[39] to labor on plantations, as have been "altered" or "changed." Describing the human onslaught on the living world in neutral terms subtly avows as fact that the natural world has no intrinsic moral standing or claim.

The impartial idiom that humanity is changing the world (not colonizing it) plays into naturalizing the human impact by portraying human action as coextensive with natural-type events. Just as ethical considerations have no relevance in the behavior of volcanoes, glaciers, asteroids, or hurricanes—which indeed do change the world, permanently or temporarily—neither do ethics apply to human behavior in relation to the nonhuman realm. Some Anthropocene analysts draw this implication explicitly: Change, they argue, occurs in nature all the time, and humanity just happens to be the natural phenomenon that is driving big changes on Earth at this particular juncture of the planet's history. According to environmental author Emma Marris, for example, since we know that nature is never static, "this means that novel [anthropogenic] ecosystems, far from being a new phenomenon, simply represent the latest changes on a dynamic Earth."[40] Journalist Fred Pearce also asserts that constant flux is a natural aspect of the world; "humans may have dramatically speeded that up, but novelty is the norm."[41] This view of humanity as "the new kid on the block," in an ever-changing biosphere, secures a spurious authenticity for the human onslaught, by implicitly presenting it as a bona fide product of the very world that is being torn apart.

Framing the human impact in a radical ecological idiom—as colonizing and destroying rather than shaping and transforming—is also, of course, a discourse. It is a discourse, however, that insists on remaining intimate with the perspectives of nonhuman existence and experience and deeply attuned to how nature's freedom has been violated by civilized behavior. Ostensibly "radical" ecological discourse stays close to the particular stories of landscapes, seascapes, and earthlings—to the *how's* of nature's so-called transformations.

For example, we could describe what European settlers did to the Great Plains as "transforming the North American prairie biome into a corn-soy-wheat-cattle anthrome." (I discuss the meaning of "anthrome" below.) But that framing bypasses the violence of that transformation, and by the same token weaves the silenced violence into human nature and into nature's order as a whole. We could also say, as the *Economist* does, that "trawlers have altered the seabed,"[42] but if we actually look at what trawlers do and have done—which is to bulldoze through three-dimensional, life-created and life-rich habitats of continental shelves and seamounts, turning them into mud and rubble— then even the word "destroy" seems inadequate; the words "devastate" and "desecrate" hew closer to the truth. What's more, we could state that fur trappers changed the Falkland Island ecology when they drove the Falkland Island wolf to extinction. But when we descend to the ground to examine how that "change" occurred, we find that the Falkland Island wolf was so friendly and curious that trappers offered the animals meat with one hand while holding a knife or club in the other, until all the animals were wiped off the face of the Earth.

These examples of the prairie, trawling, and the extermination of the Falkland Island wolf are in no way exceptional; they are typical of how civilization has carved the world. I offer them to make the point that the language of destruction, domination, colonization, and, yes, genocide too—the language describing what human supremacy looks like on the ground—stays close to the phenomena and does not shun the dirty details nor sweep them under the rug. Nor does (or should) this framing of destruction endeavor to tell another story about "human nature" as depraved and dangerous. On the contrary, it is a matter of highlighting an all-important factor: that what human beings are encultured and conditioned to believe about the more-than-human realm is decisive in how they act in and upon that world. And that the worldview of human supremacy, for way too long, has taught people that they have the prerogative to do whatever they want to the merely living, to the wild and domestic ones, to the Earth's landscapes and seascapes, and in our time to the Earth system as a whole.

Closely related to choosing a neutral idiom of "changing the world" (over the pointed language of "destroying the world"), and of preferring a depoliticized rhetoric of "human transformation" (over the political thrust of "human colonization"), Anthropocene-think regularly deploys the expression of "human-biosphere coupling" to describe the biophysical insinuation of humanity into every Earth-system variable. "The human enterprise," write Steffen and coauthors, "is now a fully coupled, interacting component of the Earth system itself."[43] The event of this "coupling"—a word encouraging an image of merging rather than one of occupation—is said to present humanity with the imperative for more scientific research; it turns out we need "a science of coupled human-biophysical systems."[44] Constituting the human impact as a natural phenomenon naturally mandates an empirical research agenda to map and understand the phenomenon—as opposed to mandating advocacy on the part of scientists and others to change it. We need to assimilate "new information commensurate with humanity's exploding capability to gather both biophysical and socioeconomic data and to analyze, interpret and model complex system dynamics," urge Steffen and colleagues.[45]

Last, mainstream Anthropocene literature naturalizes the human impact by means of introducing certain neologisms that mime ecological nomenclature to describe how humans affect the land. Chief among them is the coined term "anthrome," intended to echo the ecological term "biome" in describing large-scale conversions of nature for human purposes—especially agricultural and settlement purposes. The play of anthrome on biome directly implies that they are analogous if not similar types of phenomena.

As discussed earlier, "biome" refers to a large community of plants, animals, and other organisms interacting and affecting each other and the physical environment through relationships of symbiosis, predation, tolerance, and nutrient cycling and recycling. No biome in the natural world is dominated by any one species; rather a biome constitutes a bio-topography cocreated by innumerable interdependent beings. An "anthrome," on the other hand, is dominated by humans and their domestics, and includes other life-forms only to the extent that they are able to parasitize upon, or live in the interstices of, such human-dominated places. Thus, anthromes may include some wild creatures who are either tolerated or stealthy enough to move through them undetected. Anthromes, especially industrial agriculture landscapes and grazing rangelands, have not only been constructed through violence, cleansing the landscapes of their prior biodiversity, but require ongoing violence to sustain—the mass killing of anything labeled a weed or a pest, along with sundry bystanders in the environs or downstream.

It is possible, as I will argue later in this work, that anthromes genuinely analogous to biomes can be created in an ecological civilization. Within such anthromes, even as humans play a keystone role, landscapes will be cocreated with other species and natural processes, without constant killing being structurally instrumental, indispensable, and dominant in the process. An important example of an anthrome that is integral to an ecological civilization is producing nutritious food by intercropping a diversity of plants, in cooperation with farm animals, using organic and other eco-friendly techniques, in creative connection with wild beings and ecologies. This kind of food production agricultural writer and practitioner Fred Kirschenmann calls "food as relationship."[46] Relationship means reciprocity: neither is implicated in the least in the "anthromes" that prevail on the planet.

Landscapes constructed by means of the large-scale appropriation of Earth's net productivity, in conjunction with eradicating life-forms who are in the way, can only be called anthromes with accompanying penalties: one, of excising those characteristics of biomes, most especially biodiversity, for which no analogy holds; two, of stripping out the features of relationship and reciprocity inherent in biomes; and three, of strongly if implicitly exonerating violence as a means of making "anthromes." The way that the concept of anthrome in the Anthropocene literature exonerates violence is by means of what sociolinguists call the performative aspect of language:[47] exonerating violence is what the use of "anthrome" does by making any human mode of operation within the natural world—whether relational and reciprocal or violent and genocidal—irrelevant to the application of the label itself.

I have focused on the term "anthrome," because its coining to describe human-dominated landscapes purged of biodiversity, and to include human-dominated landscapes that are constructed and sustained by violent means (industrial agriculture being the poster case), is a significant, quasi-Orwellian conceptual-performative innovation: for by means of the explicit analogy to "biome," the notion of "anthrome" naturalizes human domination, while simultaneously obscuring the fact that biomes are in their very essence diverse, and never created much less sustained by violent means. In our time, so-called anthromes dominate three-quarters of the ice-free world while only one-quarter is wild.[48] Naturalizing human dominance by means of the term "anthrome" (and others[49]) makes this situation appear normal. Yet the status quo is completely out of step with how landscapes and seascapes come into being in the biosphere. The anthromes constructed by the dominant civilization are nature colonialism plain and simple. I believe author and farmer Wendell Berry pins colonialism precisely in defining it as an ambition to "impoverish one place in order to be extravagant in another."[50] Industrial agricultural anthromes impoverish the entire biosphere for such extravagances as feedstock, meat, biofuels, palm oil, and corn syrup—commodities that have nothing to do with enduring forms of nourishment or well-being and that are made to serve an extravagantly populous humanity to boot.

I have discussed the multivalent, mutually reinforcing ways in which

Anthropocene-think naturalizes human domination, ultimately representing the appropriation of the biosphere by a global and globalizing civilization as the latest event in Earth's natural history, an event inaugurating a new epoch to be named after ourselves. The multilevel communiqué of human takeover as a natural phenomenon can overwhelm its audience with the impression of verisimilitude. But effecting such an impression is exactly what discourse achieves by means of persistent and consistent messaging. The frame of naturalizing humanity's dominance can appear as revealing a true picture, when it is enticing its audience to stare at a mirage created by the frame itself.

Framing the human impact as natural has momentous consequences in terms of encouraging certain ways of moving forward while discouraging or marginalizing others. If human dominance is a natural history event, rather than a manifestation of the domination of nature orchestrated by a supremacist worldview, our situation does not call for a radical shift and redirection of history's course, but for reforms via techno-managerial interventions. This tunnel vision for moving forward is cemented in a sizeable slice of contemporary environmental discourse by another consistent message about human nature: that it is technologically bent and managerially oriented in its very essence.

Naturalizing humanity's impact is a form of mythmaking, in Anthropocene-think often entwined with the portrayal of the human as "the God species" and with the desire to christen a slice of geological time after Anthropos. Mythmaking is integral to the human imagination, yet this currently propounded mythology is the latest riff on the tired narrative of "the Ascent of Man." Naturalizing the disfigurement and impoverishment of the biosphere, while representing it as stemming from humanity's power to create new expressions of nature, is a move accomplished in mainstream Anthropocene literature by dehistoricizing the human and depoliticizing the human-nature power relationship.

Pleistocene Overkill Representations

In the time span between fifty thousand and ten thousand years ago, many of Earth's big animals became extinct, an event known as the Pleistocene megafaunal extinction (or late Quaternary extinction). In the 1960s, ecologist Paul Martin offered a hypothesis that attributed this extinction event to the global spread of human hunters, who encountered large prey animals unprepared for their stealth, technologies, and appetites.[51] Martin's explanation of megafaunal extirpations, called the Pleistocene overkill hypothesis, was eventually well received and is today often presented, in much popular and scientific literature, as an established explanation.

Pleistocene overkill is the last environmental discourse I dissect that represents the human impact as natural. My intent is not to assess the validity or fallaciousness of the Pleistocene overkill hypothesis. Instead, I criticize certain standard representations of it, which, openly or implicitly, convey the idea that human nature is intrinsically destructive.

A look at the research since Martin's thesis reveals that the attribution of megafaunal collapse principally or solely to human agency has been steadily disputed. Pleistocene extinction involved a complex array of conditions and causes, spanning five continents and several millennia, and including human hunting and habitat modification alongside significant (natural) climate shifts. Research findings are largely encouraging moving away from a "one-size-fits-all" account of the disappearance of the world's big Pleistocene animals. In contrast to sweeping explanations—whether anthropogenic or climatic—the Pleistocene extinction event, if ever fully understood, will involve a mosaic of causes specific to region and time, as well as to climate-related and human-related patterns around the globe.[52]

In an overview of the scientific literature on Pleistocene extinction, scientists Paul Koch and Anthony Barnosky lay out a complex picture, not only temporally and spatially, but also in the multifaceted, complicated inquiry that this extinction event requires for scientific elucidation. What counts as evidence and how it is procured include methods of timing climate shifts and human arrivals, assessments of megafaunal kill sites or their paucity, comparisons with earlier climate-driven extinctions within the Quaternary era, computer simulations, studies of paleo-ecological data regarding ecosystem regime shifts, appraisals of human population densities and hunting gear, and inferred diets of different peoples.[53] Adjudicating these variegated types of evidence is extra challenging, given the remoteness in time of the Pleistocene extinction event.

The conclusion that Koch and Barnosky reach is that "it is time to move beyond casting the Pleistocene extinction event as a simple dichotomy of climate versus humans. Human impacts were essential to precipitate the event, just as climate shifts were critical in shaping the expression and impact of the extinction in time and space."[54] In some cases, researchers argue that climate shifts might have been primary, while human hunting pressure delivered the coup de grâce.[55] Unlike human-driven extinctions in the Holocene, where the hand of human agency is indisputable, what caused the earlier disappearance of big animals on continents continues to be a matter of debate. To date, the role of humans in megafaunal extirpations in the Americas, Europe, Asia, Australia, and Africa, during the Pleistocene, is a mixed picture of being clear, contestable, synergistic, circumstantial, or still under investigation (depending on exactly where and when is under consideration).

My interest is not to weigh in on how much responsibility prehistoric humans had in the extinction of the mammoth, mastodon, saber-toothed tiger, giant sloth, dire wolf, and other formidable animals roaming the Earth thousands of years ago. I focus on ferreting out two problems with how the Pleistocene overkill hypothesis is represented, as opposed to the question of its facticity per se. The first problem is the portrayal of overkill as a sweeping and uncontested fact—especially in popular literature but to some extent in scientific literature as well. The second problem lies with an irresistible inference that Pleistocene overkill seems to invite: namely, that the anthropogenic extinction crisis is a single continuous event beginning with the human diaspora out of Africa some sixty thousand years ago and continuing all the way to this morning. I turn to a discussion of these two problems.

The problem with presenting the Pleistocene overkill hypothesis as down-and-out fact is that it oversimplifies a complex phenomenon.[56] The complexity of the Pleistocene extinction is intrinsic to the event, involving numerous convoluted facets. These include its remoteness in time, and hence fragmentary and uncertain physical evidence; divergent spatial and temporal extensions across (and within) different continents and peoples; concurrence or uncertainty of the timing of the arrival of human hunters and climatic changes; and complexity of the methodological tool kit applied to the inquiry—alongside the related imperative to make different kinds of evidence mesh into a coherent picture. Notwithstanding that Pleistocene overkill is neither indisputably nor comprehensively a "fact," unfortunately it is often presented that way. For example, the recently published and widely read "Ecomodernist Manifesto" makes the blanket assertion that "North Americans hunted most of the continent's large mammals into extinction in the late Pleistocene."[57]

Despite a complicated reality—wherein a mangle of environmental, climatic, human, and other causes in all likelihood drove continental megafaunal losses tens of millennia ago—giving prehistoric hunters the starring role has proved highly enticing since the hypothesis was first proposed. Its enticing character does not stem from the strength of evidence alone. Other factors are at play, such as the narrative appeal and simplicity of Pleistocene overkill. As Stephen Wroe and his colleagues put it, "the concept of rapid global extinction of large animals, solely through violent human activity, evokes sensational imagery and its mass appeal is strengthened by its simplicity."[58] In other words, not only does Pleistocene overkill provide a bottom-line explanation, it contributes an alluringly dramatic one to boot.

Playing into its narrative appeal is the mythic imagery of a big-brained, technology-wielding, super-social, carnivorous primate, who, even in the Stone Age, was capable of making an oversized mark on the biosphere. Thus, beyond the sensational and Occam's razor attractiveness of the deadly hunter story, the allure of the Pleistocene overkill hypothesis derives from its tacitly placing humanity's planetary impact right in the lap of "human nature." It seems to nicely rest the case about the big, wicked picture of *Homo sapiens'* sojourn on Earth from deep time to the present moment.

That brings me to the second representational problem with Pleistocene overkill: namely, the embedding of its depiction in narratives that portray the human-driven extinction crisis as one long event beginning tens of millennia ago and ongoing in the present. "A massive extinction event has been underway for some 40,000 years"[59] is how, in one fell swoop, the extinction crisis is rendered. Thus, the simplified portrayal of Pleistocene extinction as solely human-caused finds correspondence in the general portrayal of anthropogenic extinction as a single event. Anthropologists Todd Braje and Jon Erlandson, in a paper subtitled "Late Pleistocene, Holocene, and Anthropocene Continuum," convey the idea: "The wave of catastrophic plant and animal extinctions that began with the late Quaternary megafauna of Australia, Europe, and the Americas has continued to accelerate since the industrial revolution."[60]

This kind of representation takes a more boldfaced shape in popular literature, where unsavory implications are more comfortably drawn: "A bad smell of extinction follows *Homo sapiens* around the world," writes historian and author Ronald Wright.[61] Later he adds: "The last tree. The last mammoth. The last dodo. And soon perhaps the last fish and the last gorilla. On the basis of what police call 'form,' we are serial killers beyond reason."[62]

This meshing of distinct extinction episodes (Pleistocene and Holocene, continental and island, prehistorical and postindustrial) into one unified storyline is what I will call "the extinction continuum." It is common if not ubiquitous in the environmental literature. The extinction continuum lumps events that had different causes, combinations of causes, and degrees of confidence about causes (and combinations of causes) into an all-encompassing picture. This picture tends to be highly appealing to those who suspect a "serial killer" hidden within human nature; at the same time, this encompassing picture reinforces this serial-killer view. Yet it is fair to counter that the extinction continuum, as such, is not supported by empirical evidence, for the domain of empirical evidence involves specific, fine-resolution geographical and temporal contexts. Sweeping portrayals of divergent extinction events run roughshod over important discrepancies, nuances, and gaps of knowledge. The extinction continuum is a narrative, one analogous to the projection of a completed puzzle picture from fragmentary puzzle pieces: the continuum story fills in the "blanks" by activating the assumption that exterminating life is what big-brained, technology-wielding humans naturally do.

The monolithic extinction-continuum storyline elides significant considerations that undermine it. It is especially troubling in the representation of continental and island extinctions as similar "natural kinds." (The last mammoth. The last dodo.) Human-driven island extinction, however, "says little about the likelihood of similar events on continents."[63] There is incontrovertible evidence for anthropogenic extinctions on big and small islands throughout the Holocene (for example, Mauritius, Hawaii, Madagascar, and New Zealand). Indeed, islands are highly vulnerable to both direct and indirect human impacts. On the other hand, the role of early humans in mainland extinctions is complicated by concurrent climatic and ecological shifts and by the fact that animals would eventually learn that human hunters are dangerous and, in seeking to avoid them, could more readily escape them on large landmasses.

The extinction continuum does not only mesh island and continental extinctions, but it also lumps Pleistocene and Holocene impacts, which are qualitatively distinct. Even if prehistoric hunters *did* play a lead role in the extermination of Pleistocene megafauna, it does not follow that this event was of a similar ilk with the ecological extirpations and species extinctions humanity has been causing in the post-agricultural phase through the present.

If meat-eating humans, wielding relatively sophisticated hunting gear, took out the mammoths, the woolly rhinos, the big camels, and the rest, such extinctions were unintended consequences—the kind of high impact that nonnative (invasive) species can have. Following the first (possibly catastrophic) contacts of prehistoric settlers and native biota, a new dynamic equilibrium emerged between recently arrived humans and their still biodiverse environs. On the other hand, the relationship that supervened between humans and nonhuman nature *after* animal and crop agriculture was established, and the historical sprint of empires began to unfold, was entirely another thing. We can hardly call the exterminations that have dogged civilization's heels "unintended consequences." Consider, for example, how a nineteenth-century buffalo hunter couched his hunting work in accordance with his culture's vision: "The passing of both the Indian and the buffalo was inevitable. The great development of the West could never have begun until their occupancy ended."[64]

The civilizational turn inaugurated an intentional and systematic assault on the natural world, setting into motion the takeover of the biosphere via stop-start, incremental, or blitzkrieg operations. With the entrenchment of civilization, ideas that demeaned nonhumans and wild nature, elevated the ("superior") human, and constructed the distinguished human identity that we have inherited became established over time. In accruing fashion, supremacist ideas were ensconced within conventional actions, linguistic formulations, commonsense beliefs, institutional arrangements, and amoral technologies all bent toward, and synergistically upholding, the invasion of the natural world for its conversion into being for people: the global ecumene.

This perspective on the impact of humanity within the Holocene epoch as a single continuous event opens itself to the same charge of a sweeping assessment. Admittedly, this view blankets over the watershed transition into the modern era. The latter immensely escalated impact, what with the plunder of the Americas in "the long sixteenth century" (1450–1750),[65] the rise of fossil-fuel energy, the establishment of capitalism and its growth and opulence imperatives, the mechanized transportation revolution, the population explosion after humanity reached the one billion mark, the industrial fishing onslaught, and the post-1950s "Great Acceleration."[66] In the last five hundred years, humanity's blow on the biosphere has increased by orders of magnitude in comparison to what occurred previously in the Holocene.

Yet the supremacist worldview that has been handed down and across cultures, undergirding humanity's mode of operation during the entire historical period, remains the fundamental engine keeping the human enterprise moving full throttle forward. The undead big story of human superiority and entitlement has been decisive. This is so because humans are not, first and foremost, beings carried onward by inertia, or duped by power-hungry CEOs or other elites, or somnambulating along plugged into gizmos and gadgets—as true as these diagnoses are to some extent or other. Rather, human beings live according to and immersed in meaning, and the inherited worldview of human supremacy is the shared belief system from which elucidations of "who we are" and "what we do" are sourced. It is the dominant mythos of human distinction and prerogative that permits humanity to continue multiplying its economic connections, turning a blind eye to population growth, spreading technological infrastructures, emptying the ocean, burning down the rain forests, damming the rivers, installing more and more factory farms, and tearing up the crust of the Earth. The grand narrative of human ascendancy and its "peak experience" today of human planetary ownership breathe meaning into the expansionist enterprise, imbuing its every extension with legitimacy if not necessity.

Against Naturalizing the Human Impact

The prevalent belief that human nature is to blame for the ecological crisis offers a seductively simple explanation that discourages deeper thinking into what is driving the biosphere's predicament. The view that planetary dominance has supervened from human nature ("naturalizing the human impact") is itself a socioculturally propagated narrative in the broader culture and in a good deal of environmental thought. This storyline is empowered by the diverse ways it is expounded, as I have illustrated with the three prominent discourses examined in this chapter. Yet naturalizing the human impact also gains force by tapping into a deeper vein: it streamlines with the longstanding Differential Imperative predilection to regard the human as naturally—and obviously!—distinguished from all other life-forms.

Naturalizing the human impact on the biosphere echoes, and simultaneously derives power from, the enduring motif of the Differential Imperative. Their logics are seamless: Humanity has arrived at the end point of planetary ascendancy precisely because humans are essentially different from the rest of life. This line of reasoning appears as a nobrainer. What is less obvious yet begs contemplation, however, is that by clinging to the Differential Imperative—and the worldview of human supremacy it eventually spawned—humanity precisely produced the ideational and material engine of nature's takeover that has led, incrementally and (more recently) in leaps and bounds, to the present situation. Thus, instead of unreflectively buying into the "naturalness" of the human difference, it is more revealing to consider the myriad ways the identity of human distinction is trained into the human and thereby continuously performed and reproduced.

Immersion into the view of the human as a distinguished entity begins at childhood, continues throughout human life, and is reinforced in *all* forms of cultural and historical learning. In observing the thorough training through which humans come to see and know themselves as special, we begin to glimpse how that identity is socioculturally constructed at such a fundamental level that it becomes what phenomenologists call *constitutive* of the human. To say that human distinction is a constructed identity means that human beings are coached to believe it, come to embrace the belief in it, and proceed "to perform in the mode of belief"[67] in it by virtue of total immersion in a sociohistorically produced way of life: a way of life that considers nonhumans as secondary (if not disposable) Earth denizens and treats their homes as steal-able without malfeasance; a way of life that dictates the reproduction of subjugating actions that have been inherited, institutionalized, and normalized; and a way of life that bestows elevated meaning upon itself, through linguistic constructs and loaded dualisms that contain and broadcast the trope of human distinction. Human supremacy is a social achievement (if that is the right word) not a biological inheritance.

The historical entrenchment of human supremacy has not been expressive of some core human identity or inborn makeup, but the upshot of the nonstop discursive and pragmatic performance of that worldview. What's more, by cumulatively revamping the biosphere, the human-supremacist worldview has succeeded in masking its character as a sociocultural order, appearing instead as objectively confirmed in the mirror of the world's transformation. Human conditioning is so saturated at every level that it becomes invisible as a mode of training, and its missive of superiority and entitlement appears instead to be borne out of "the facts of the matter." A world of human-dominated surroundings—and their representational abstractions in, for example, geopolitical maps—has itself been produced through the operations of an entity that regards itself as special and privileged. The physical materialization of the supremacist worldview, in turn, gets reflected back to the human mind reinforcing the worldview itself. This is what it means to say that human distinction has become constitutive of the human. It appears entirely obvious: It is a "compelling illusion,"[68] since the dominant culture assiduously produces, as a matter of course, the distinguished human identity qua natural entity. To assume that this "natural entity" is the culprit behind the ecological crisis is an easy—another obvious—next step.

Human supremacy reproduces itself via the intergenerational sociocultural crafting of the distinguished human. Because this training (along with its reinforcements) is airtight, human distinction subsequently appears as naturally so, and its reality as a sociohistorical project is, borrowing the words of Judith Butler, thoroughly "obscured by the credibility of its own production."[69] The formation of the human takes place within a field of power—of human totalitarian rule over Earth—that erases its historical and political character by presenting "the distinguished human" as ontologically given. This platform magic of the dominant culture, trailing behind it the historical weight of millennia, is responsible for creating the "foundationalist fiction"[70] of Earth as the Planet of the Humans.

To subvert the compelling illusion of human specialness and the foundationalist fiction of Earth as humanity's property calls for arduous efforts in deprogramming at personal and community levels, cultivating new ways of seeing, questioning normalcy, performing alternative types of communities within the biosphere, engaging in ecological restoration projects, and generally venturing in practice and imagination into new possibilities of being human in the world. Such efforts will "eventually bring a qualitatively new pattern into being and into consciousness, so the new paradigm is more apt to be discovered than to be legislated."[71] Breaking out of our supremacist brainwashing requires of us, right now, "to cultivate a kind of exile from the comfort of assured truths."[72] Discovering life after human supremacy will involve protracted work carried forward by future people.

Human-nonhuman power inequality is the historically erected and accreted foundation of elevating the human, the original political act, an imposed apartheid that has been disguised as the natural ground upon which politics per se are vaunted as having their domain proper in "the complex world of human affairs." There exists no such natural ground—it is politics all the way down. Otherwise put, that disguised-as-natural ground must be demolished so that an integral human way of being and living may emerge.

To interrogate the deep-seated belief in human specialness is not a revolution in thought: it is not about coming to realize that we are *not* special, after all, nor is it about ploddingly teasing out the ways we are special and the ways that we are not. All such musings stay ensconced inside the box of the Differential Imperative—by either striving to negate difference or endeavoring to qualify it with appropriate nuance. A far more compelling possibility of being and action lies in the wisdom of what philosopher Matthew Calarco calls the stance of *indistinction*, which he elucidates as arising from "the desire to inhabit the world from perspectives other than those of the classically human subject and to explore the passions and potentials that are found in such spaces of encounter."[73] We must "seek a radical exit," he urges, from "the violent and denigrating logic of anthropocentrism and its associated metaphysics, institutions, economy, and discourses."[74]

The new adventure of the human in the biosphere and the emergence of another human identity get under way the moment the question itself—"How are humans different?"—is dropped. This heralds a revolution in being. For when the human finally sheds the compulsion of the identity of distinction, the entire world (and existence) will appear in a completely new light. The world will appear in a light truer to what is, because it will no longer need to be filtered through the lens of human specialness in order to conform to that perspective. The world will open up to as yet barely charted possibilities of perception and experience and will blaze in a light that self-illumines it more closely to what it is. A mystery. Or, in the preferred understanding of indigenous people, a gift. Yet even such handsome words remain only words. Opening toward the world as what it is, without having to lay it on the Procrustean bed that reconfigures it as beneath and for humanity, will open the same unbounded space within the human. Breaking out of the cage of supercilious solipsism, what kind of winged human will emerge? In setting the world free of the human-supremacist worldview, we free ourselves as well from its petty, mind-numbing tyranny.

This particular turning turns on intimacy. The etymological roots of "intimacy" lie in the Latin *intimus*, meaning "inmost, innermost, and deepest." Later, intimacy morphed into meaning "closely acquainted, very familiar." Thus, within the meaning of intimacy lies that which journeys inward to the innermost and that which simultaneously journeys outward to the very familiar. Likewise, human intimacy with Earth will journey the human into its inmost being; perhaps "who we are" will then become intimated. Intimacy with the Earth will, at the same time, journey the human toward becoming closely acquainted with every place and every being of this planet; this existential nearness can only be sustained through a care so great that it borders on tenderness. When intimacy turns into action, it becomes the verb "to intimate," meaning "to make known." Among the things that will become known is that whenever a hierarchical imperative of difference is brought into operation, all possibilities of intimacy are foreclosed. (A kind of shattering occurs that manifests as trauma, psychic pain, separation, loneliness, and rehearsals of violence.) As long as humanity remains tethered to the belief system of superiority and entitlement, the human cannot come near the plenum of being and participate; the very belief in human difference closes the door. When we simply drop that belief, however, without needing to revise it or replace it, a revolution in being will occur (occurs). We can call that revolution coming home, or we can call it becoming intimate.

As long as the belief is perpetuated that the ecological crisis is caused by "human nature"—precisely because human nature is essentially different—humanity stays locked into the view of human ascendancy as "naturally so." In these reifications of human identity and human ascendancy lurks grave danger. It is the danger of the human-supremacist worldview swallowing up both stage and protagonist: turning the stage from a cosmos into mere materiality and turning the human away from the nearness of intimacy into the crass and superficial modality of user.

In closing, my grievances with "naturalizing the human impact" do not intend to imply that there is no such thing as human nature (i.e., that humans are born a blank slate), nor that there exists a pristine human nature in innate harmony with human and nonhuman others that we must seek to recover. The will to dominate exists within us in seed form, affiliated perhaps with life's imperative for self-preservation. What the anthropocentric worldview has concertedly done is to till and fertilize that particular seed, growing it into a monstrous and now global superstructure of domination. Thus, the drive of the sociohistorical conditioning of the human to dominate the nonhuman realm did not come out of nowhere: it came, in some sense, out of our nature. Yet it is entirely incorrect to understand that drive as coextensive with who we are.

Who we are is children of a living world. This universal knowing and its tropes of participation, reciprocity, gratitude, belonging, and love reside deeply within us. We can choose to cultivate and grow them into a beautiful human way of life.

FIVE

The Trouble with Debunking Wilderness

In beauty I walk.
With beauty before me, I walk.
With beauty behind me, I walk.
With beauty below me, I walk.
With beauty above me, I walk.
With beauty all around me, I walk.
It is finished in beauty.
It is finished in beauty.
It is finished in beauty.
It is finished in beauty.

NAVAJO SONG

Alongside empirical findings and conceptual developments that foreground humanity's shattering global presence, in the past two decades the status of wilderness has also undergone sustained scrutiny. It has become standard fare to claim that wilderness no longer exists and may well have been gone for millennia. As the *Economist* sums up this stock verdict, "wilderness, for good or ill, is increasingly irrelevant."[1] In this chapter, I examine a second discursive knot: the trend to label the idea of wilderness a flawed empirical referent and misguiding lodestar. I argue that the debunking of wilderness itself begs to be debunked, for it lends the human domination of nature considerable support and blocks the imaginary of an alternative human inhabitation from surfacing into awareness.

Etymologically echoing self-willed (autonomous) nature, "wilderness" refers to the reality and conception of places in the natural world that remain free from exploitation, conversion, and exterminations by civilized humans. (Indigenous people, on the other hand, who have lived in reciprocal integration with the natural world have been, and can continue to be, integral members of the wilderness.) As the primal manifestation of nature, wilderness contains its most biodiverse, fecund, creative, and dynamic expressions. To dispense with the referential vehicle for this manifestation of the natural world—as a concept both for the real world and the human imagination of that world—is equivalent to jettisoning the original blueprint of nature. This jettisoning, in turn, amounts to doing away with the conceptual vehicle that lucidly reflects the diminishment and destruction effected by limitless human expansionism. To discard the concept of wilderness, as certain strands of environmental thought propose, is to forgo a contrasting baseline reality that reveals the full picture of civilization's onslaught.

More than mere semantics is at stake here. Dispensing with the idea of wilderness edges humanity toward the slippery slope of "anything goes" regarding the natural world. What's more, doing away with the notion of wilderness undermines solidarity with the freedom of nature and its denizens, and vitiates censuring incursions into wild nature driven by human entitlement. By resisting such rote statements as "wilderness is gone," "wilderness has been gone for a long time," or "wilderness is a sociocultural construct," reengagement with the promise that wilderness holds out now and for the future becomes possible.

It was in the 1990s that the concept of wilderness—defined as nature "pure and unsullied" by any human presence—came under fire from academics in the fields of geography, social ecology, and environmental history. Since then the debunking of wilderness also found its way into popular culture and the environmental movement. In the last twenty years or so, wilderness has regularly been assailed as myth and cult, labels that speak to an offensive against both the reality and the idea of wild nature.[2]

Four lines of argument against wilderness are typically advanced. One is that pristine nature no longer exists, for humanity has touched every bit of the biosphere—if not directly then indirectly via the global effects of climate change and of toxic and long-lasting chemicals found in the most remote locations. The second argument against wilderness is that nature untouched by people has not existed for a very long time; humans have been shaping the natural world for millennia—certainly since its postagriculture phase but even stretching into prehistory (as previously discussed in connection to the Pleistocene overkill hypothesis). The third antiwilderness argument is that the idea of wilderness is objectionable in implying that people and nature are ontologically separate entities; but, it is countered, humans themselves are part of nature. The fourth antiwilderness argument is that wilderness is a thoroughgoing cultural construct—and a modern American one at that.

These multiple charges leveled at the tenability of "wilderness" appear for many to cinch the case against it, while those who cling to the wilderness ideal are deemed misguided or "consumed by nostalgia." "Having failed to grasp the historical extent of humanity's ecological influence, they practice a wilderness-worship that's not only ineffective but delusional."[3]

As a unified package, antiwilderness arguments have done an extraordinary disservice to the biosphere. Their corrosive effect has been to puncture a spirited defense of nature's freedom, wild beings, and inherent processes. As a whole—the first two arguments denying the reality of wilderness and the last two gutting the idea—they fog human thought so profoundly as to hamper the ability to distinguish between humanity colonizing nature versus humanity being a part of nature. "Not only is wild nature being destroyed in physical reality," writes ecopsychologist David Kidner, but "its obliteration is completed by its elimination from history and imagination, undermining our sense of ourselves as embodied creatures with a natural past, so that a felt resonance with the natural order gives way to a largely cognitive assimilation into the industrial symbolic order."[4]

Indeed, the academically initiated damage to the reality and idea of wilderness has the dubious distinction of having shored up the ruination of wilderness that occurred since those first discrediting analyses began to be circulated. Since 1993, a vast amount of wilderness—twice the size of Alaska—has been destroyed or seriously degraded around the globe. "Environmental policies are failing the world's vanishing wildernesses," observes biologist William Laurance. "Despite being strongholds for imperiled biodiversity, regulating local climates, and sustaining many indigenous communities, wilderness areas are vanishing before our eye."[5] *Vanishing before our eyes*—while in the meantime an influential portion of the contemporary academe had been teaching students that wilderness is an American cultural fiction.

The denial of the primordial reality of wilderness, along with the falsification of the concept, tie profoundly into naturalizing the human impact and thus normalizing nature's domination, as previously discussed. For defending wilderness—that is, defending nature's self-creation free from constant revamping, using, and exploitation—is the core node of resistance against the natural world's domination. Once wilderness is out of the way as reality and idea, human dominance looks more like "prominence," which, in turn, looks like just another phase of Earth's natural history. The original ontology of the biosphere as a vast biodiverse wilderness, revealing—through contrast—how human domination obliterates life's richness, disappears from both world and mind.

I argue for the profound need for environmental thought to untie this discursive knot. We need to eschew recent (mostly academic) anti-wilderness platitudes and to rethink wild nature—to awaken clear sight into the meaning of nature's freedom and creativity, and to inspire the arising of a global social movement that will defend these qualities and seek to reinstate the reign of wilderness in the biosphere. With that intent, I turn to respond to each of the four arguments against wilderness sketched above.

Argument 1: Pristine Nature Is No More

All environmentally informed people, and increasingly the broader public, know that no landscapes or seascapes remain unaffected by humanity. Defining wilderness as 100 percent pristine only serves to set the stage for knocking its reality down and rejecting the usefulness of the idea once and for all. It is also a convenient definition for developers, for it swings the doors open to additional impact with the excuse that nothing untouched remains anyway.[6] The 100 percent pristine definition of wilderness is spurious. In our time, wilderness has a more robust and pragmatic meaning: it consists in extensive natural areas that are mostly undisturbed by civilized human activities; many such regions remain in the world, and more can be restored.[7] Wilderness protection does not amount to an illusion of safeguarding pristine places, but, in the words of environmental author Paul Kingsnorth, involves defending "large-scale, functioning ecosystems worth getting out of bed to protect from destruction."[8]

Wilderness areas encompass lands and waters sizeable enough to support wide-ranging species, especially big carnivores and herbivores who need largely people-free, expansive spaces to live, disperse, and migrate. In the words of conservation biologists Michael Soulé and Reed Noss, large carnivores, "require extensive, connected, relatively unaltered, heterogeneous habitat to maintain population viability."[9] This requirement extends beyond the needs of big animals. "Ecologists and conservation biologists have known for decades that small isolated parks leak species. Smaller populations have smaller gene pools in which maladapted traits are more likely to become fixed. Smaller populations are more vulnerable to drought, pests, hard winters or simple bad luck."[10] Along with large size, another critical feature of wilderness is contiguity—nature unfragmented by road systems, industrial agriculture, grazing rangelands, human settlements, mining operations, or other developments. Within wildlands and wild waters, free ecologies are seamlessly blended.

There are two interrelated reasons that the wilderness features of bigness and contiguity are so essential. First, these features allow bountiful and diverse populations of plants, animals, and other organisms to flourish, to move (in response to changing conditions and/or in accordance with their natures), and to evolve into new life-forms. Evolution is tangibly unfolding in wild landscapes and seascapes, because in wilderness members of a species can be numerous and often extravagantly so; in wilderness, species are composed of distinct populations in different locations (metapopulations); and, in wilderness species can slowly diversify into distinct varieties and subspecies, which constitute incipient manifestations of speciation (the birth of new species). Second, the features of big and unfragmented allow wilderness to withstand disturbances—whether natural or human driven—such as wildfires, extreme weather events, or volcanic eruptions. For example, many millions of acres of forested lands are needed to absorb one raging wildfire without life in those lands becoming completely devastated.

In short, wilderness is the matrix within which more and new life emerges and within which life is sheltered and sustained. Preserving wild nature has nothing to do with the misunderstanding of freeze-framing some "original" static and ahistorical set of species and processes—and keeping that state intact in perpetuity. "A more sophisticated concept of wilderness preservation"—and the only correct one—"aims rather to perpetuate the integrity of evolutionary and ecological processes," explains environmental philosopher Holmes Rolston.[11] Thus, wilderness preservation is precisely about sustaining change in its most creative expressions: wilderness ensures flourishing ecological and evolutionary dynamics, because big, connected, and free nature is the unrivaled author and keeper of such dynamics at their most exuberant and intricate levels.

Big, connected, and free is real-world wilderness, sometimes called "ecological wilderness" as Noss points out.[12] Wilderness refers to self-willed places that are whole enough to unfurl themselves in life am-plifications and entanglements that are self-sustaining and generative. One hundred percent unblemished by human influence is not a requi-site criterion for nature to count as wilderness. While everywhere in the world wilderness has become "an endangered geographical species,"[13] as journalist Brandon Keim states the case, wilderness areas still ex-ist. What is equally important is that biodiversity-rich wilderness can be restored, enlarged, and reconnected—in a word, rewilded—both on land and in the seas.[14] "The future of biodiversity," write ecologists Paul Ehrlich and Robert Pringle, "is not just what we can save of what is left, but also what we can create from what is left."[15]

A robust understanding of wilderness is indispensable for wilderness restoration, increasingly called "rewilding" by conservation scientists and practitioners around the world. Wilderness—biodiverse terrestrial and marine places that should remain beyond the clutches of civi-lized humans—is an empirical reality and pragmatic starting point for rewilding initiatives. The term "rewilding" was coined by environmen-talist Dave Foreman, while decades later author George Monbiot de-scribes it as a massive restoration of the natural world, which includes removing infrastructures like roads, fences, and so on; reintroducing extirpated species where appropriate; and returning to wild nature its will and ability to self-heal.[16] The reduction and eventual cessation of all management activities is a key component of the rewilding vision. In Ned Hettinger's words, "turning nature loose to head off in a trajec-tory that we do not specify is key to rewilding."[17] As the Wild Europe Initiative puts it, "a naturally functioning landscape that can sustain itself into the future without active human management is the ulti-mate goal of the rewilding approach."[18]

Environmental thinker David Johns defines rewilding as "creating a system of core areas, corridors, and buffers," in order to support "all native species, all ecosystem types and processes, disturbance regimes, and resilience."[19] Resilience means safeguarding the conditions for wild nature's dynamism, wherein, even as constant flux is present, life's qualities of abundance, complexity, and diversification sustain them-selves. "Cores" refer to natural areas that can be designated as pro-tected or are already protected and can be enlarged, while "corridors" connect the cores. Connecting wild landscapes that are fragmented is the most effective way to save species and preserve ecologies.[20] "Buf-fers" describes places abutting rewilded landscapes, which people use lightly and in ecologically friendly ways. A quintessential example of a buffer is small-scale, diversified, organic farms that benefit from the proximity of wild nature (via the presence of healthy hydrological conditions or pollinators, for example), while simultaneously interfacing with wilderness in wholesome, biodiverse-in-their-own-right patterns.

Large-scale protection, restoration, and connectivity are critical to rewilding and to its constituent dimensions of cores, corridors, and buffers. "To protect Earth's living membrane, the biosphere," states nature writer Julia Whitty, "we must put nature's shattered pieces back together. Only megapreserves . . . hold that promise."[21] Anthropogenic climate change has only added urgency to the larger rewilding mandate, for many species challenged by shifting climatic conditions and/ or extreme weather patterns "will need big gene pools to draw from and lots of different places to which they can move."[22] Rewilding calls for limiting humanity's demographic, economic, and infrastructural expansionism, while at the same time restoring the natural world at continental and oceanic scales.

Argument 2: Humans Have Been Modifying Nature for Millennia

The second argument that critics make against wilderness is that human beings have been shaping the natural world for millennia. This argument runs together two separate lines of evidence that need to be teased apart. One line of evidence of long-standing impact is that our species' systematic land-altering activities began with the emergence of agriculture some eight thousand years ago: eradicating, converting, and marginalizing wilderness has thus been long under way.

Another line of evidence (of the second argument) is that people *everywhere*—including the indigenous societies of the Americas and elsewhere—left sizeable ecological footprints on the land and humanized large areas of nature. Thus, native people's hunting and cultivation practices, elaborate settlements, trading ventures, and fire use, among other activities, cast the existence of "wilderness," even among the indigenous, into grave doubt. I address the two lines of evidence of the second antiwilderness argument in turn.

Argument 2a: Civilization's Millennia-Long Impact on Wilderness

The long history of civilization's impact on the natural world is indisputable. For example, looking at the cradles of Western civilization, we find landscapes that are almost wholly anthropogenic and profoundly degraded. Mediterranean forests, woodlands, and scrublands are among the most critically endangered regions of the world.[23] Forest cover in the Middle East and the Mediterranean basin was extensive in the early Holocene. Deforestation began millennia ago, literally and figuratively fueling the march of empires. The adage that "forests precede civilization and deserts dog its heels" has its origins in the human-driven desertification of these regions.[24] Relentless deforestation has thus been one of civilization's oldest, nonstop, and still ongoing nature-destroying practices; it chewed up temperate forests and has turned today, additionally, to the tropical and boreal zones.

The systematic killing and persecution of wild animals was also coextensive with the emergence of civilization, especially due to their habitats turned into grazing ranges and cultivated fields. Wild animals also suffered tremendous targeted blows. To mention one notorious episode of sustained cruelty, Roman trade of wild animals for their gladiator shows and other public games—which occurred across all the empire's amphitheaters and not only in Rome's famous Coliseum—was as thorough as it was brutal. The Romans captured, displayed, and slaughtered whatever "exotic" animals they could get hold of, from wherever they could reach: lions, leopards, tigers, aurochs, elephants, hyenas, camels, hippos, giraffes, crocodiles, ostriches, rhinos, and more. All told, "the total number of animals killed as a result of Roman games is staggering."[25] Over the centuries of Rome's rule, the result of this bloodbath was the ecological extirpation and regional extinction of wild animals everywhere Roman trade extended.[26] The mass slaying of animals in Roman amphitheaters persisted until the sixth century AD.[27]

As environmental historian Keith Thomas succinctly put it, "human civilization indeed was synonymous with the conquest of nature."[28] Historian Roderick Nash offers a similar synopsis: "For thousands of years the success of civilization seemed to mandate the destruction of wild places, wild animals, and wild peoples. The game plan was to break their wills."[29] Historical-ecological scholarship revealing civilization's protracted impact is critical, but so is how we interpret the revealed destruction. One thread of environmental analysis interprets civilization's long history of conquest as displaying an inborn human proclivity toward appropriating the natural world for agricultural development, breeding practices, pasture conversion, energy production, material-culture development, trade expansion, and other aggrandizing social purposes. Nature's takeover is decoded as revealing a hardwired feature of humans as managers, engineers, natural-limit breakers, tamers of life, and (in our time) planetary masters. Environmental scientist Erle Ellis, for example, voices the lure of this interpretation: "We must embrace our history as ancestral shapers and stewards of the biosphere," he argues.[30]

The historical gaze can yield a different interpretation: that civilization has been a thoroughly mixed enterprise, which along with its legacy of "positive goods" has also trailed a dark side that has "besmirched and bloodied every page of history."[31] The ascent and hegemony of the human-supremacist worldview constitutes the dark side of civilization. Taking onboard the long history of its consequences can serve to awaken the aspiration for a new civilization that supersedes the pathology of supremacy. Historical understanding, in other words, can rouse us to change the course of history away from its nature-colonizing ideology and inertia, rather than yielding to its march. Recognizing that civilization has been destroying wilderness for millennia does not carry the implication that we must abandon the reality and idea of wilderness. Instead, we can move toward creating an ecological civilization that will thrive within a rewilded biosphere.

Argument 2b: Indigenous People Shape the World, Too

The fact that civilization has been repurposing wild nature for millennia is one line of evidence against the tenability of wilderness offered by its critics. The other is that indigenous people have done likewise, also for millennia. Indeed, wilderness critics regard "the pristine myth"[32] as having special cachet in the Edenic image of pre-Columbian North America. As an antidote to the ostensibly mythical notion of North America as a once untouched wilderness, it has become fashionable to emphasize that the continent was pervasively inhabited and shaped by native people.

Throughout the New World, there were settlements, roads, trading outposts, constructed mounds, and farmed lands, as well as the signs of fire use and hunting, well before European settlers arrived. (See figure 2 for a pictorial illustration of this "indigenous humanization" perspective.) Thus, wilderness detractors submit, when we compare the pre- and post-Columbian New World, the difference we find is not one between pristine nature, on the one hand, and humanized landscapes, on the other: what we find is "different forms of [land] management," as author Emma Marris puts it.[33]

The recent emphasis on avowing human modification of all manner of landscapes runs roughshod over divergent relationships between different peoples and nature, by making human modification—no mat-

Figure 2   Indigenous people shape the world. (Source: Getty Images.)

ter what it looks like, how it is effected, or what its extent—the Main Noteworthy Event. "The problem with lumping all landscapes together as anthropogenic," author Richard Manning observes, "is it makes no distinction between a soybean field in Iowa and a fire-created meadow in the Bob Marshall Wilderness."[34] It can similarly yield a failure to distinguish between Disney World and Yellowstone National Park.[35] The obsession with finding an anthropogenic imprint on the land is so overriding that preposterous claims—notably, the Amazonian rain forest being "largely anthropogenic in form and composition" or "largely a human artifact"—have been put forth in print.[36] For anyone who has ever leafed through a large-format rain forest book, held spellbound by nature's masterpieces of living beings, the news that Amazonia is ("largely") human handiwork will surely surprise.

There never was a pre-Columbian "Garden of Eden," it is averred regarding the New World, only a "cultural narrative of one."[37] If the view of the New World as pristine wilderness is a cultural narrative, the revisionist view of the New World as humanized is only another—but this one to the tenth degree. The story of North America as shaped and managed by its native human inhabitants commits the proverbial error of missing the forest for the trees. When Europeans landed on its shores that continent was a vast wilderness bursting in all its seams with biodiversity. The choice to foreground its plentiful and dynamic human inhabitants skirts and consistently downplays "extravagant accounts of a heavily fruited land, rivers full of fish, rich soils, and abounding animal life" recorded in the 1600s.[38] The thousand-year-old giant redwood and sequoia trees of the Pacific Northwest are legendary, but the white pines of eastern North America when European settlers arrived were also practically sequoia-size.[39] What life scientists call "complete food webs" crisscrossed the continent. The classic food-chain pyramid, from soil microbiota to large carnivores, was complete and thriving. The great numbers, diverse populations, and distinct subspecies of wild animals—to focus on animals as an example, though the same was clearly the case for plants, fungi, and other organisms—reveal that the kaleidoscope of evolutionary play was spinning throughout the animal kingdom.

North America was wall-to-wall carpeted by an estimated 250,000–500,000 wolves (three or more subspecies), well over the *world* wolf population today.[40] A near continent-wide range was also the case for cougars (also known as mountain lions and by other names).[41] A continental range almost certainly applied to elk, who consisted in six subspecies, and numbered around ten million when Europeans showed up.[42] (Two elk subspecies were driven to extinction, while elk numbers were decimated and their historic range hugely contracted.) Roaming the western part of the continent were an estimated thirty-five million pronghorn antelopes consisting in four subspecies.[43] About fifty thousand grizzlies also ranged over the west, from California to the mid-plains and from central Mexico to Alaska.[44] As for the most iconic of the continent's animals, the American bison, their numbers were between twenty and thirty million or more—constituting "the largest aggregation of large mammals on Earth."[45] They consisted of at least two subspecies, and lived from the Gulf of Mexico to Alaska and almost coast to coast. The woodland caribou historically ranged across the forests of North America, from Maine to Washington State; the species is now critically endangered, with the last US herd reduced to about forty animals, most of them living inside the Canadian border.[46]

Then there was Aldo Leopold's "biological storm," the passenger pigeons, whose dung would coat thousands of acres, flight would block the sun for hours, and forest roosting extravaganzas would bring mighty branches crashing down. "Yearly the feathered tempest roared up, down, and across the continent," wrote Leopold, "sucking up the laden fruits of forest and prairie, burning them in a traveling blast of life."[47] Passenger pigeons numbered in the billions.[48] In pre-Columbian times, beavers ranged over most of the North American continent. According to naturalist Ernest Thompson Seton's estimates, beavers numbered between sixty and four hundred million.[49] Beavers created habitat for countless species. There was an abundance of fish once migrating between land and seas on both coasts—salmon, alewife, shad, sturgeon, eel, and others. Fishing for sturgeon in the Chesapeake Bay, for example, was compared to gathering vegetables from a garden.[50] The history of what happened to the vast numbers of Atlantic and Pacific coastal life after the Columbian landfall—whales, sea otters, seals, walruses, fish, sea turtles, seabirds—is a story, to invoke a fitting saying, to make angels weep.[51]

There is a vociferous, highly politicized debate about how many native people originally lived in the Western Hemisphere, with proposed numbers ranging between forty and one hundred million. For North America, estimates span between 1.8 million and 18 million—diverging by an order of magnitude, as author Charles Mann has pointed out.[52] Those who argue that the pre-Columbian landscape was extensively modified by people tend to be partial to higher estimates of American Indians before they were decimated by Europeans, most especially by introduced diseases after first contact.

Let us concede that there were *many* people in pre-Columbian North America, and add another numbers game to the mix: Imagine the aggregate number of wild animals before the Old World colonizers proceeded to decimate them. (Beavers alone well outnumbered even the highest end of the spectrum of estimated people in North America.) Recall also that I have made no mention countless other fauna: of white deer, mule deer, and moose; reptiles and amphibians; black bears, wolverines, bobcats, martins, and lynx; the mesopredators; and the continent's abundance of birds. It is safe to venture that North America's pre-Columbian "wild vertebrate zoomass" easily dwarfed its "anthropomass."

Wild nonhumans were not only bountiful in numbers, they played critical roles in shaping ecologies on continental and oceanic scales, what with their wetlands engineering, their vast migrations, their nutrient cycling across the continent and between land and seas, and other ecological roles. Indeed, humans are not the only species involved in fire ecology. Large herbivores—whose biomass was stupendous before European settlement—play a significant role even in fire regimes. "By altering the quantity and distribution of fuel supplies," write ecologist

William Ripple and colleagues, "herbivores can shape the frequency, intensity, and spatial distribution of fire across a landscape."[53] Thus, it was not just people substantially molding nature; it was also non-humans, even more substantially, doing the same. Acknowledging the abundant biodiversity of pre-Columbian North America, while recognizing also that the continent was widely inhabited by people, should not invite the conclusion that wilderness did not exist, but the hopeful realization that humanity and wilderness can beautifully coexist. Indeed, human modification was far from the "Main Noteworthy Event" of North American landscapes, except for observers who wear one pair of glasses for looking at the world—made by Anthropocentric Lenses, Ltd. A clear, panoramic view of the pre-Columbian North America, of its full house of diverse living beings, reveals the continent's cocreation by all its inhabitants, nonhuman and human.

North America was prodigious in wild beings and ecologies, co-inhabited by native people, for a combination of reasons: the *comparatively* low numbers of human inhabitants, the nonexistence of a Native American "livestock industry," and, most importantly, the animist cosmologies of Native Americans, which brimmed with respect for the natural world.[54] As indigenous author and activist Winona LaDuke writes, "according to our way of living and our way of looking at the world, most of the world is animate." "Looking at the world and seeing that most things are alive," she continues, "we have come to believe, based on this perception, that they have spirit. They have standing on their own."[55] Animism, ecopsychologist Ralph Metzner explains, "sees all life-forms, including animals, plants, rocks, forests, rivers, mountains, fields, seas, winds, as well as sun, moon, stars, and the total cosmos, as pervaded by and interconnected with spiritual energy and intelligence."[56] Animist vision does not only see spirit pervading both animate and inanimate creation, but also recognizes "that the world is full of persons, only some of whom are human."[57] The animism of indigenous peoples is thus far more sophisticated than contemporary philosophical quibbles about what matters most—wholes or parts, species or individuals, habitats or organisms—as if such compartmentalizing schemes actually capture the nature of the living world, or can order our priorities for responding to and caring for it.

It is neither an idealized nor a romantic understanding that native people held nonhumans and their homelands in high regard. Tellingly, indigenous stories and practices always blend portions of place, non-human nature, and peoples (or tribes), emphasizing balance, relationship, and identity between them. In the words of LaDuke again, "indigenous people have taken great care to fashion their societies in accordance with natural law."[58] That means, among other things, observing and respecting natural cycles and "taking only what you need and leaving the rest."[59] In indigenous cultures, writes author Stephanie Mills, "venerating particular animals or plants as totems or regarding certain places as sacred informed the human sense of landscape and set ritual limits on the exploitation of those beings."[60] Of course, Native Americans had an impact on their environments. But living in accordance with tenets of gratitude, reciprocity, and respect, they did not assimilate the continent wholesale nor wantonly slaughter and persecute nonhumans to annihilation—at least, not before the conquerors poisoned their minds with greed and/or debilitated them to such a degree where they were left no choice.[61] In their thousands of years of inhabitation, indigenous people never came remotely close to the mass destruction wielded, in a handful of centuries, by a people conditioned by a supremacist culture.

To what extent native people shaped the land with fire is also disputed: but no matter what the exact degree, fire use was clearly not practiced at the expense of a thriving wilderness. Early twentieth-century environmental thinker and activist Bob Marshall articulated poignantly what was legible on pre-Columbian landscapes and seascapes: "The philosophy that progress is proportional to the amount of alteration imposed upon nature never seems to have occurred to the Indians."[62] In depicting Indians as thus restrained in their relationship with the natural world, Marshall was expressing admiration for the aboriginal people of North America.

Today, however, new storylines are circulating about Native Americans, which, instead of emphasizing their animist cosmologies and related restraint, highlight how they, like everyone else, were also "managers of their natural resources." In Charles Mann's *1491*, for example, we learn that many "Indians were superbly active land managers—they did not live lightly on the land."[63] The popular press has been quick to catch on to the implications of this framing regarding the human impact more generally. What Mann "is interested in showing us," a *New York Times* review of *1491* elaborated, "is how American Indians—*like all other human beings*—were intensely involved in shaping the world they lived in."[64] According to the same article, to regard Native Americans in any other way would be "dehumanizing."[65] In response to this in-vogue revisionism of Native American lifeways, we might exclaim: O tempora! O mores! To think that, at long last, indigenous people have been rehumanized through revelations of their underappreciated managerial ways!

The weird stereotype inversion of native people, from romanticized "noble savages" to pragmaticized "resource managers," has become customary in recent years.[66] Here is environmental historian Ted Steinberg's spin: "To see the Indians as the continent's 'first environmentalists,' living in harmony with the natural world until the Europeans set foot on the land and destroyed it, is a view that, at worst, is demeaning to Native Americans. It turns them into savages incapable of making aggressive use of the environment and thus unworthy of any rights to the land in the first place."[67] A number of strange assumptions are tightly piled into this passage: including, that to call indigenous people "the first environmentalists" would demean them; and that to appreciate that they actually practiced a nonaggressive relationship with nature would imply they were too primitive to exercise the capacity of being destructive.

Such newfangled ideas about indigenous people have not corrected erstwhile erring, idealized views of them. The swing from noble savage to resource manager has substituted one ideology-laden view with another. What's more, when we consider the words and ceremonies of Native Americans themselves, historically recorded as well as preserved in contemporary expressions, the noble-savage stereotype arguably hews closer to the truth. "Indigenous people of [the Northwest] forests, and all over the world," writes indigenous author Robin Kimmerer, "offer traditional prayers of thanksgiving which acknowledge the roles of fish and trees, sun and rain, in the well-being of the world. Each being with whom our lives are intertwined is named and thanked."[68] The Navajo song cited at the start of this chapter is not exactly what one would affiliate with a "natural resource management culture." Native American scholars and activists have noted that the relationship between their ancestors and the more-than-human world was one of love (not management), one of respect (not resourcism), and one of communion and solidarity (not dominion). Indigenous literature on human-nonhuman relations emphasizes kinship, reciprocity, limits, and gratitude.[69]

To recap, in the long history of civilization's impact and the nature-shaping activities of indigenous people, antiwilderness arguments impute a uniform pattern on the human: that our nature is to tame wilderness, transgress biophysical limits, harness technologies for sculpting nature, and generally manage landscapes large and small. An inclination to see an anthropogenic imprint on the land is so overriding that enormous differences between the impacts of civilized and indigenous people are glossed over as just variations in degree. For example, environmental authors Michael Schellenberger and Ted

Nordhaus write: "The difference between the new ecological crises and the ways in which humans and even pre-humans have shaped non-human nature for tens of thousands of years is one of scope and scale, not kind."[70] Along with such transhistorical and transcultural conflations, we witness a profound blurring of distinct worldviews; people everywhere turn out to be, or to have been, land shapers and resource managers. "Millennia of exuberant burning shaped the Plains into vast buffalo farms," opines Mann, projecting his own culture's mindset onto American Indians.[71] The largest pre-Columbian biome of North America was, according to the same author, "a prodigious game farm."[72] The inability to see through such distorting projections, and to blithely ignore the baggage such descriptions carry, is as mind-boggling as it is disconcerting.

For such a straightjacket depiction of all humans to have any credibility whatsoever, the importance of worldviews—especially the rift between human-supremacist ideas and animist cosmologies—in cultivating divergent relationships between peoples and the natural world must be systematically discounted. A fallacious understanding of the human condition is subtly propounded: that cultural narratives of origins and identity, as well as shared myths or religious ideas about human-nonhuman relationships, float disconnected over some deep-seated, pragmatically bent human predilection to mold and lord over all else.

Painting the human connection with the natural world with one broad brush—of humans as aggressive nature shapers—peddles the dangerous perspective that it is inexorably in us to invade, take over, convert, kill, and (of course) manage the natural world at will. The inference drawn from this view is both inevitable and pernicious: "Native Americans managed the continent as they saw fit. Modern nations must do the same."[73] The line between the limited and respectful indigenous impacts of the past—effected within cultural contexts of reciprocity with the nonhuman realm—and the terraforming scale of civilization's entitled and demolishing activities is not just blurred but erased. The human takeover of the biosphere is turned into a destining that we are tacitly urged to accept as ordained.

Argument 3: The Idea of Wilderness Implies Human Separation from Nature

It is critical to challenge this blunt tack of universalizing humanity's domineering strut—tightly coupled with the debunking of wilderness

—for it is a storyline that insinuates that our present-day dominance in the biosphere is destined. Accepting the supremacist narrative of human ascendancy has the perilous effect of locking us into one view of the human, as though it were cross-culturally universal and inescapable. It conveys the idea that the best we can do is to become "good managers of natural resources," while saving some seminatural places here and there as "natural capital" for provisioning "ecosystem services." It prods us to forget that human beings have manifested enormously divergent stances toward the natural world.

The two arguments discussed thus far—that wilderness no longer exists and that wilderness has not existed for a very long time—advance the perspective that wild nature is a long-defunct *reality*. That however is not where the debunking ends. Additionally, wilderness is claimed to be a bankrupt *idea* for implying an ontological separation between humans and nature. On the contrary, wilderness critics affirm, humans are part and parcel of the natural world and not separate from it.

In response to this third line of wilderness critique, it must be emphatically stated that wilderness does indeed include a substantial degree of separation between humanity and the more-than-human world: this is a virtue of the wilderness idea not a flaw. Affirming the need for some separation between the human and nonhuman realms reminds us that people do not need to live and extract from everywhere, nor bring technologies and infrastructures to bear on all places. When we enter wilderness we enter a living space where other life-forms, nonhuman processes, and nature's freedom have the upper hand. Wilderness is the geographical dimension where ecological and evolutionary processes unfold with minimum interference and where wild beings are free to carry on as they will.

The human-nature boundary that the idea of wilderness espouses— of terrestrial and marine realms that we visit with profound discretion and at our own risk—embodies an ethos of respect and not the meaning of a sharp ontological divide. Boundaries are appropriate and necessary components in all healthy relationships—both relationships between people and relationships between people and the nonhuman world. Complying with such boundaries is not only an expression of esteem: it is a prerequisite of intimacy. The price of violating boundaries, on the other hand, is rupture and trauma. Civilization's running down of the biosphere—its disregard of any boundaries—does not signify that civilized humans have been part of the natural world, but on the contrary signifies their presumption to do violence to wild nature.

It is precisely because civilization has had no sense of appropriate boundaries—no etiquette toward the nonhuman world to speak of—that the analogy of rape, abrasive though it is, has had purchase in expressions of environmental rage. While no one would accede a rapist's defense that he was seeking oneness with his victim, when wilderness advocates condemn the civilized assault on the wild they are charged with ignoring the fact that humans are just a component part of the more-than-human world.

Rather than foregrounding the reality that occupying wild nature does not remotely signal human unity with it, wilderness debunkers slip in an interpretation of civilized humanity's boundary violations as demonstrating human inseparability from the natural world. Thus are people befuddled into confounding the colonization of nature with our oneness with it.

The idea of wilderness as including a human-nature boundary, observed from respect, is an ethical concept and forms a nucleus of resistance on behalf of what remains of the wild against its full-blown assimilation. "We lost the wild bit by bit for ten thousand years and forgave each loss and then forgot," writes author Jack Turner. "Now we face the final loss."[74] Defense against the final loss of wild nature is motivated by respect and love, and not by the alleged perspective that there exists some "essential dichotomy" between people and wild nature. It was civilized humans who enshrined a hierarchical dichotomy—and the natural world's assimilation by the human enterprise is precisely how *dichotomy* has been operationalized.

A human-nonhuman hierarchical regime, conjured by civilized humanity, has undergirded the self-consigned prerogative to eradicate, use, and convert the wild. The mainstay of the wilderness idea and wilderness activism has been conscientious objection to this plunder and to the human-nonhuman constructed hierarchy that grounds it.

Argument 4: Wilderness Is an American Idea

When settler William Bradford arrived, in 1620, at the Northeast shores of the American continent, what he saw from the bow of the Mayflower was "a hideous and desolate wilderness full of beasts and wild men."[75] He was not voicing his personal opinion, but expressing the perspective of his culture. It turned out also that his words were prophetic. In the short centuries that followed and to this very day, the

American wilderness was overhauled and harnessed, and "beasts and wild men" were exterminated or confined to assigned plots of land.

The history of what transpired in the New World involved a complex array of appurtenances that the Europeans brought with them, incidentally or purposefully: diseases, a livestock plague, a market economy, the concept of private property, technological means (especially after the Industrial Revolution), a "breathtakingly anthropocentric"[76] reading of their main religious text, and an inveterate supremacist mindset and vision—for the men and women who made landfall on the New World had "a highly developed sense of their own racial and ethnic superiority."[77] Upping the ante of such tremendously formative forces, population growth joined the fray with the eventual mass movement of people across ocean and continent. In the nineteenth century alone, the settler population climbed from roughly five million to seventy-six million people.[78] By "God's providence," and with Anthropos as his instrument, the continent was to be turned into a neo-Europe.

It was not the anthropocentric metamorphosis of land and seas that was anything new. The Europeans of the Old World, and their Mesopotamian, Egyptian, Greek, and Roman predecessors, were long practiced in wilderness conquest. What was new was the speed. It was as if the entire history of human empire, of its mode of operation of cognitive belittlement coupled with geographical takeover, was condensed into a sped-up fireball of itself. "The image of the wilderness east of Mississippi," writes historian Richard Slotkin, "changes from 'desert' to 'Garden' in a century and a half, while that of the Great Plains exhibits a similar change in less than half that time—from its purchase in 1803 to the realization of its economic potential before the Civil War."[79] In short order, whatever could be chewed up for power, profit, and subsistence was: fish; whale and seal oil; fur, pelts, and feathers; timber; gold and silver; oil, coal, and (eventually) natural gas; and land and soil. "The process by which [the colonists] came to feel an emotional title to the land," continues Slotkin further down, "was charged with a passionate and aspiring violence."[80]

One would not think that the words "passionate and aspiring" are good modifiers for "violence," but Slotkin's assertion is emically accurate. For the white settlers, wild nature was not a place that (virulent and rapacious) violence could be inflicted; it was a place that held the potential for remolding and improvement—for the purpose of which violence was merely expedient.

In 1843, an anonymous author of an article titled "Taste and Fashion" articulated the colonists' guiding paradigm of the human-nonhuman binary with verve. "Nature builds no house or temple, spins no dress," he or she wrote, continuing:

She writes no poetry, composes no music, presents us with no forms of intercourse [!]. Having given out forms enough to beget activity in human taste, she scants her work that we may go on and exert a creative fancy for ourselves. The wild woods are cleared away, the green slopes are dressed and laid out smiling in the sun, the hills and valleys are adorned with beautiful structures, the skins of wild beasts are laid aside for robes of silk or wool. In a word, architecture, gardening, music, dress, chaste and elegant manners—all inventions of human taste—are added to the rudimental beauty of the world, and it shines forth, as having undergone a second creation at the hand of man.[81]

This description of what was done to the continent and its inhabitants is sanitized, twisted, and exuding the characteristic sickly stupor of supremacist reasoning. It was a massacre. Wild animals of land and seas were turned into commodities or exterminated with abandon. Common Old World tags for forests included wild, dreadful, gloomy, desert, uncouth, melancholy, unpeopled, and beast haunted.[82] No wonder European colonists promptly razed the New World's forests: the tracts of New England, New York, and the entire Eastern coast; the expansive woodlands of the Great Lakes region; the longleaf pine forest that stretched from Virginia south to Florida and west to Texas; the old-growth floodplain woods of the South; and the towering forests of the Pacific Northwest. Clear, swift-moving streams and rivers, created and sheltered by forests, turned muddy, slow-moving, and impoverished of life. Mining operations transmogrified landscapes into moonscapes and despoiled ecologies.[83] Wherever the land could be turned over to crop cultivation and livestock grazing that was effected—with passionate and aspiring violence.

The violence was not only physical, but conceptual as well. Wherever the colonists settled, they tended to impart arbitrary names from the places they came from or their celebrities. (For example, Virginia, where I currently reside, was named after England's Queen Elizabeth I, who was thought to be a virgin.) Such naming, though dissociated and insolent, was at least sociopsychologically understandable. The most drastic conceptual violence done to the continent was the imposition of "the Grid" (passed into law in 1785), which stamped an abstract, homogenizing, and fictional (i.e., ungrounded in reality) measure on the land, thereby forcing its transformation into what the grid-image summoned.

Its checkerboard pattern divided the land into six-mile-square boxes, further subdivided into one-mile-square boxes, and then divided again into quarter sections of 160 acres each.[84] This standardizing scheme—superimposed without any regard for natural features, contours, or inhabitants of places—sought "to bring wildness under the governance of a monoculture of rational control."[85] The grid was, as historian David Nye describes it, a totalizing system.[86] It was a system designed to turn the land into occupied land, a system that demanded and quickened occupation, and a system that gave the impression that to fill it in was, in Nye's insightful words, "an automatic historical process."[87]

Looking back at the effected physical and conceptual violence it is probably fair to say that most settlers saw no wrongdoing. What people saw (if they could even look up from their daily grind) was a necessary, sanguine, and futuristic march of what the nineteenth-century "human enterprise" would crystallize as "Progress." Most settlers understood their history in process through the lens of what Nye calls "the master narrative of the second creation," a narrative that even after it became "indefensible," as he puts it, "Americans remain loath to abandon. . . . It has ceased to be merely a story, and become a national account of origins that confers entitlement to the continent to white immigrants."[88]

Even as the majority of immigrants remained complaisant about that entitlement, a minority emerged whom the experience of colonizing the American continent jolted into awakening. In one way or another, those people abdicated from the dominant culture: whether they became philosophers like Ralph Emerson, critics like Henry Thoreau, preachers like John Muir, organizers like Bob Marshall, analysts like George Perkins Marsh, or poets like Robert Frost, Walt Whitman, and Emily Dickinson, they broke rank from their society and blazed a different path. This was the kind of "American" (and their countless anonymous followers and activists) who changed the concept of wilderness and birthed the wilderness movement.

Their awakening instigated the reconceptualization of wilderness in a way unheard of in Western culture and agitated for protecting wild nature from its wholesale subjection to what Robert Frost called, in his lovely poem "Rose Pogonias," "the general mowing":

A saturated meadow.
Sun-shaped and jewel-small
A circle scarcely wider
Than the trees around were tall;

Where winds were quite excluded,
And the air was stifling sweet
With the breath of many flowers,
A temple in the heat.

There we bowed us in the burning,
As the Sun's right worship is,
To pick where none would miss them
A thousand orchises;
For though the grass was scattered,
yet every second spear
Seemed tipped with wings of color,
That tinged the atmosphere.

We raised a simple prayer
Before we left the spot.
That in the general mowing
That place might be forgot;
Or if not all so favored,
Obtain such grace of hours,
that none would mow the grass there
While so confused with flowers.

To call wilderness "an American idea" is both spot-on and entirely incorrect. It was an idea born of the American experience of witnessing the rehearsal of human empire's violence in blitzkrieg mode. It was not an idea that profoundly converted the mainstream, which today continues to war against the natural world. In this very hour, the American establishment is giving over wholesale landscapes to poisonous fracking operations;[89] gridding the Gulf of Mexico and Alaskan coasts into blocks auctioned off to oil companies for offshore drilling;[90] exporting (and pipeline laying for) climate-wrecking fossil fuels; killing carnivores with relish;[91] subsidizing corporate timber extraction from the national forests; and repurposing (and drenching in herbicides) vast swaths of land and freshwater for the major products of mainstream America—meat, dairy, corn syrup, and ethanol. Wherever on the North American continent wild places and creatures have been saved from annihilation, it is because committed people have gone to great lengths to battle for them.[92]

There is, however, a larger universal victory of the wilderness idea: its conceptual transformation, by a cadre of clearheaded people and their minority reports, from referring to a hideous and desolate world to a beautiful and fecund one. This was an accomplishment as rare as it is auguring, an *actual* deconstruction and reconstruction of a nature concept from a purely negative referent to one redolent with abundance and stature—henceforth one akin with Walt Whitman's ecstatic cry of "containing multitudes."

Every human culture has gifts that eventually become the universal heritage of humanity. The gift of wilderness birthed from the American experience was a new way of seeing, which has set into motion a new way of being. For the transformation of the concept did not only provide a shield against wild nature's complete assimilation into the ecumene, but also prefigured (as I will elaborate in the last chapter) the possibility of another kind of human inhabitation altogether. Herein, indeed, lies the most nefarious aspect of wilderness debunking: its failure to appreciate an extraordinary Western cultural-conceptual transfiguration, which sprung from witnessing violence for exactly what it is. Not passionate and aspiring, but virulent and rapacious.

The Viability of Wilderness

Toward the end of his famous essay "Walking," Henry Thoreau wrote: "In short, all good things are wild and free."[93] He paired the words "wild" and "free" indelibly, so that their conjunction has become virtually a household expression. It is not entirely correct, though. It is less "wild *and* free" than it is "wild *is* free." Life's essence is uncaged. Unhampered, life moves, propagates, diversifies, and populates the planet, while creating and filling worlds within the planet. There are trials of strength in nature but nothing like jail bars. There are natural barriers but no structures akin to cement walls, barbed wire, enforced monocultures, and miles of drift nets and longlines notched with millions of hooks. There are elemental dangers of all kinds, but no snares, machine guns, poison 1080, pesticides, or habitat-bulldozing trawlers. Slicing the topography are rivers, mountains, and ravines, but no prisons of nature cut off and insulated from one another like the ones asphalt, industrial agriculture, mountaintop mining, or tar-sands extraction construct.

Wild is rooted in will-ed. Will is inside wild. Freedom is inside will, because will rebels when its freedom is hampered or stolen. Once I saw (what was probably) a jaguarundi in a cage barely bigger than his body. This was in the backyard of a tourist shop outside of Cancun, Mexico, in 1985. He sat like a sphinx. I approached slowly. Like a living statue of dignity, he neither looked at me nor dignified my presence with as much as a twitch of a whisker. What happens when a human being encounters such sordid and debased shackling of living beauty? When freedom and goodness are caged by some supremely arbitrary and ugly act? An inner tempest of rage and grief, and an etching of memory that never fades. The human spirit is forever allied with what is beautiful and good and free and wild, and will never stop marching toward that destination as the abode for human life. Beyond all doubt, this return to Earth will occur in the future. Why wait that long when so much is dying and disappearing right now?

Wilderness is the self-arising ground made of innumerable and diverse beings interacting with one another and with the elements. Wilderness contains the "chaos that gives birth to a dancing star."[94] The human is one of the dancing stars it birthed. Wild nature is the Creative, the unbounded space within which the unexpected arises and the highly patterned recurs. Wilderness, "rich with liberty,"[95] painted the living canvas from which the highest value—freedom—was directly perceived by the human mind.

SIX

Freedom, Entitlement, and the Fate of the Nonhuman World

All the animals, the plants, the minerals, even other kinds of men, are being broken down and reassembled every day, to preserve an elite few, who are the loudest to theorize on freedom, but the least free of all.

THOMAS PYNCHON

Sea turtles have roamed the ocean and coastal seas for over one hundred million years. Of the seven species of sea turtles in existence today, six, if not all, of them are threatened with extinction, ranging in conservation status from critically endangered to vulnerable.[1] In pre-Columbian times, it is estimated that there were over one billion sea turtles in the ocean. Those numbers are down by at least 95 percent.[2] Ferdinand Columbus, describing an encounter in the Caribbean, recorded that within "twenty leagues, the sea was thick with turtles so numerous it seemed the ships would run aground in them and were as if bathing in them."[3] There were fifty to one hundred million green sea turtles (alone) in the Caribbean.[4] Over the course of four centuries, New World voyagers killed millions of green and hawksbill turtles for consumption and commerce.[5] In just the last fifty years, all the Caribbean species have declined in numbers by 99 percent.[6] With their populations decimated, they have become functionally extinct, no longer playing the important roles they once did as cocreators of their habitats.

Sea turtles everywhere face the same threats, for the onslaught against them is systemic. The first line of attack comes from industrial fishing, especially from the gill-net fishing, longline fishing, and shrimp trawling that have yearly killed hundreds of thousands of turtles. The second line of attack is persecution of the animals themselves and raiding of their nests. Despite their endangered status, turtle eggs continue to be taken and sea turtles continue to be killed for their alleged medicinal and aphrodisiac properties, for the products made from their bodies and shells, and for their meat.

The third line of attack on sea turtles comes from pollution. Half the leatherback sea turtle corpses sampled around the world were found to have plastic in their stomachs. What's more, coastal pollution is favoring the spread of an emergent disease among sea turtles that is causing deadly tumors.[7] The fourth line of attack stems from development, which has taken away most of their nesting grounds. And housing and tourist development threatens even those nesting beaches that remain: light pollution disorients hatchlings who die heading in the wrong direction; and human settlements attract scavengers such as raccoons, who snatch up sea turtle eggs and newborns. The fifth line of attack on the horizon is climate change. Where there is no development lined up behind their nesting beaches, sea turtle habitat may have a chance to move inland and thus survive the rising seas. Where there is development, however, the nesting beaches will go under.

That sea turtles survive despite the decimation of their numbers and the continued onslaught on their being is testimony to life's strength and to the heroic efforts that citizens and scientists around the globe have been making to rescue them. In the long run, however, neither life's fortitude nor labors of love can save creatures from the systemic assault they are experiencing. The only thing that will allow sea turtles not only to continue existing but to thrive again is to return to them their freedom to be who they are and live as they will.

The predicament of sea turtles stands as an example of what is happening to so many earthlings, with precipitous drops in populations from multiple pressures, intractable obstacles facing migratory and wide-ranging species, human-impact-related diseases spreading, and animal starvations on the rise as the human juggernaut deprives them of their dwellings and livelihoods. Poet and deep ecologist Gary Snyder warned decades ago that "creatures who have traveled with us through the ages are now apparently doomed, as their habitat—and the old, old habitat of humans—falls before the slow-motion explosion of expanding world economies."[8] Since he wrote those words in the early 1970s, the ecological crisis has morphed into a planetary tide of catastrophes. Even so, human expansionism—of world economies and global trade, human numbers, and industrial infrastructures—marches on unfazed.

Expansionism as Enhancing Human Freedoms

Understanding how the natural world is coming undone from the tsunami of growth is understanding a lot, but not enough. Shoring human expansionism is its shared perception as a normal modality. Its normalization is so entrenched that proposals to contract the human project—to reduce the human population, degrow the global economy, and massively pull back agricultural, extractive, and infrastructural impositions on the natural world—appear fringe to the mainstream. In fact, such proposals tend to be filtered out of the mainstream before they can even be appraised, one way or another.

At a deeper layer yet, the normalization of expansionism and the recoiling against any proposal to contract the human presence are underwritten by the tacit belief that the nonhuman realm does not have the intrinsic standing nor merit the moral consideration to mandate designing human life in such a way as to not harm that realm. The demotion of nonhuman being is rarely voiced openly, for the very articulation of a supremacist belief tends to unmask and undermine it. Rather, the decimation and impending demise of sea turtles and so many other beings is witnessed (or not) as an inexorable "change" that has come into their world and is impinging upon their existence.

Along with the layers of assumptions normalizing growth and dooming creatures who have traveled with us through the ages, another formidable obstacle—the third discursive knot I consider—has emerged in our time. What is fueling the prerogative to keep expanding, and the unthinkability of contracting, is a prevailing ideological framing: that human expansionism is bringing more degrees of freedom to more and more people. In this chapter, I tackle this particular knot wherein the ongoing growth of the human enterprise becomes conflated with, and celebrated for, enhancing people's freedoms.

I am not referring to freedoms affiliated with human rights, such as the right to one's religious and cultural beliefs or to equal treatment regardless of race, gender, ethnicity, sexual orientation, or physical ability. These hard-won and still-fought-for freedoms are not exercised at the expense of the natural world. Indeed, the freedoms enabled by human-rights victories are an ideological and institutional forerunner for the capacity to recognize, and the grounds to grant, analogous freedoms to nonhumans and places. The fight for human rights augurs the possibility of a "historical continuity between different social emancipator struggles and the quest for the liberation of nature."[9] In environmental historian Roderick Nash's words, the "natural rights tradition, which has extended the moral community in the past to include African Americans, native peoples, and women, should now turn to the task of liberating another oppressed minority: Nature."[10] Since human rights are nearly universally valued, there is a real potential for the liberation of nonhuman nature to become a universal ideal, as well.

In this context, I am using the idea of freedom in its more mundane meanings. The dictionary is useful for elaborating. According to *Merriam-Webster*, qualities of "being free" include the absence of necessity, coercion, or constraint in choice or action; liberation from restraint; independence; and/or the quality of being released from something onerous.[11] Today, people seek to enlarge the scope of their freedoms within the biosphere exactly along those lines: by more or less taking from and acting upon the natural world as they will; by increasing the physical means and geographical scope of human mobility; by enlarging the range of places where food production and settlements can be located; by augmenting the infrastructural networks for transporting huge amounts of commodities and raw materials; by multiplying the means of convenience, from a throwaway economy to burgeoning household appliances; and by proliferating the nodes of communication to ease the flow of virtual stuff and to expand cyber-connectivity between people. As historian Dipesh Chakrabarty has noted, most modern freedoms are energy intensive, standing "on an ever-expanding base of fossil-fuel use."[12]

The freedoms to live or go where people please, take and do as desired, boost the exchange of the quantity and variety of stuff, consume a diversity of foods at any season, enjoy the time and effort released by modern conveniences, and extend the scope and speed of the means of mobility and communication—all these freedoms are extolled and largely experienced as broadening human horizons. They enable the enjoyment of life experiences that come with personal mobility and travel, choice between a broad gamut of commodities, consumption of diverse goods, cross-cultural exchanges, long-distance communications, and the smooth flow of information, data, entertainment, and money. The modern lifestyle offers, in a nutshell, an embarrassment of riches, wrapped in the mantle of enhancing freedom—delivering more options, more possibilities, and more experiences to increasingly more people.

This situation, however, trails a shadow: as modern freedoms are increasingly disseminated to billions of people, so do the freedoms of uncountable nonhumans become obliterated or constricted. The inverse relationship between human and nonhuman freedoms can be highlighted by borrowing from *Merriam-Webster* again: the more people come to enjoy the *absence* of necessity, coercion, or constraint in their choices and actions, the more is the *presence* of necessity, coercion, or constraint imposed on the choices and actions of nonhumans (see figure 3). "Just as the other side of the coin of the ancient democracy was slavery," observed political theorist John Rodman, "so the other side of the coin of modern human freedom has been the domination and exploitation of external nature."[13]

It is crucial to acknowledge, of course, that none of the aforementioned human freedoms are (yet) available to all people. But they are swiftly becoming available to more and more people, and they are nearly universally valued, so much so that these freedoms are tagged for equal distribution—a social mission already under way. Critical theorist Lewis Mumford noted decades ago that there is broad agreement across the entire political spectrum that "every member of society should have a share in its goods,"[14] and today those goods include the virtually limitless human freedoms within the globe. These freedoms

Figure 3   The inverse relationship between human and nonhuman freedom. (Source: Graeme Chapman.)

are what social theorists call hegemonic: while currently enjoyed by a segment of humanity, they appear to be in demand across its entire spectrum. Thus, Mumford called the establishment's proffered goods "a magnificent bribe"; they appear to be virtually irresistible.[15]

Because the freedoms of the modern lifestyle are highly prized two consequences follow. The first consequence, as I just mentioned, is the imperative to spread them since all people should have the right to enjoy them. The second consequence is an inclination to deny, downplay, or remain unaware of the dark side of these freedoms—that they require and rest upon an offensive against the natural world and nonhuman autonomy. Since the sought-for freedoms are incoherently premised on eradicating others' freedoms, this inner contradiction inevitably calls forth the concealment of its shadow. The social collective shields itself from the knowledge that broadening the gamut of human choices and actions entails extinctions, eradication of wilderness, unnecessary death and suffering, nonstop killing and persecution of wild animals, the constriction of nonhuman movement, and the loss of magnificent phenomena like animal migrations.

The web of life is unraveled and restitched into a downgraded pattern, in order to enable unrestricted human experience of movement, access, use, consumption, and connectivity. Avoiding the knowledge that these privileges are founded on eradicating nonhuman freedoms arguably underlies the present-day cultural invisibility of biodiversity collapse and of the Sixth Extinction. Mass extinctions are extremely rare and catastrophic events, but the human-driven one is seldom deemed newsworthy by the dominant media. This is not an incidental oversight: the destruction of life's diversity must remain invisible or obscure in the public domain, because it is fully entangled with the freedoms that people (are goaded to) value and seek.

The social mission to disseminate modern freedoms, coupled with the imperative to elide that they are beholden to colonizing the biosphere, is demolishing life's variety, abundance, and complexity, while miring the human mind in a conflation of freedom with entitlement—and preventing the realization of authentic freedom that can only be all inclusive.

Social Structural Supports

The freedoms described above are high ideals of modernity and sanctified in such concepts as "the freedom of the seas," "the free market," and "free trade," as well as "the commonwealth," which refers to the wealth of the natural world for people to unlock and share (rather than the world for all earthlings to enjoy). The dissemination of freedoms within the biosphere to take and do as desired, escalate the mobility and geographical scope for all things human, enlarge the choice between a flood of commodities, magnify the consumption of processed foods and beverages, sustain the nonstop flow of raw materials and cyber-stuff, and amplify global connectivity all require considerable and concerted social structural support. Indeed, modern hegemonic freedoms are secured by means of linguistic, technological, and institutional social structures, which work together reinforcing and supporting one another.

Linguistic structures embed the conceptual dimension of representing the world as human owned. The notion of "commonwealth" is one example, but there are many others such as ecosystem services, natural capital, working landscapes, fisheries, and the like. The most widely used and generic of human-possessive concepts is that of "natural resources," which constitutes (as discussed earlier) the master concept out of which spin-offs are generated. Thus, concepts such as natural capital, fisheries, livestock, timber, and so forth already incorporate the idea of resources: they are concrete instantiations of that concept. Portraying the natural world as human owned immensely facilitates human freedoms within it by means of conceptually constituting it as a rightful field for any human activity.

Technologies are designed to impinge upon that human-owned world in order to appropriate its wealth, contour its geomorphology, and crisscross it with infrastructural networks. The modern technological arsenal has three characteristics directly relevant to subjugating the natural world: it is extremely powerful; it is constantly upgraded and augmented; and it is constructed with no ethical deliberation whatsoever regarding the nonhuman realm. We can think of industrial trawlers, mountaintop-removal machinery, hydrological fracturing operations, offshore drilling rigs, or continent-spanning latticeworks of highways and pipelines as examples of how the technological dimension contributes to provisioning food, energy, and mobility for people via the ruthless subdual of the more-than-human world and catastrophic expansionism across seas and land.

Last, institutional structures include economic, political, and legal arrangements such as government subsidies and policies, international treaties, and financial bodies that enable and expedite the invasion of geographical space, bankroll agricultural, fishing, logging, and mining operations, and ensure the relatively smooth flow of what is extracted or manufactured.

Sociologist Max Weber called the phenomenon of converging and resonating social spheres their "elective affinity."[16] The elective affinity of linguistic, technological, and institutional structures pertaining to securing access to nature's plenum is a most powerful force. On the one hand, it entrenches a regime of nature domination, which "limits autonomy and imposes a style of living over which [nonhumans] have [no] choice."[17] On the other hand, it assembles a consensual reality wherein the social collective regards and experiences that regime of domination—produced by the elective affinity of "legitimate" social structures—as the real world. The regime solidifies into an enduring condition, because of the congruent junction of commonplace conceptual renderings, constantly produced and upgraded technological assemblies, and relatively long-lived economic, political, and legal institutions. The same triangulation of language plus technologies plus institutions gives rise to consensual reality, by overwhelming human consciousness with the certainty that dominance over the natural world is a sound state of affairs. As a consequence, destructive activities like industrial fishing, endless highway and road construction, damming and rerouting of rivers, continental-scale applications of herbicides, and the like, not only harden and endure, but are also regarded by the collective as valid ventures in the real world.

Freedom of the Seas

The ocean is a straightforward case to see how the elective affinity of the language, technology, and institutions of industrial fishing has coagulated into a regime of plunder, while at the same time building a consensual reality within which industrial fishing appears as a normal way of acting in the biosphere.

What is done to the seas is profoundly linked to the fact that fish are branded a resource for harvesting. When, for example, the well-intentioned author of a *Science* article warned that "the harvest of marine resources [is] now at or past its peak,"[18] there was widespread failure to notice that this state of affairs is in large part due to the very way it is described. The real world of fish becomes all but condemned through their constitution as a "resource for harvesting," or as previously discussed, "fisheries" and "fish stock." These conventional monikers appear to refer to something real in the world, when in actuality their work is to label the animals for mass consumption. The concepts of fisheries, stock, or marine resources name fish in such a way as to consistently broadcast into the collective mind stream that fish belong to people. So when we read that "Antarctic krill is one of the world's most underexploited marine stocks,"[19] we can anticipate the fate awaiting those "underexploited stocks"—as well as the penguins, seals, and whales who eat them.[20]

While conceptual renderings make mass killing seem unremarkable and prefigure the killing fields, words do not do the killing. Language must ally with a technological arsenal, as well as with economic, political, and legal institutions, to deliver the full metal jacket of human expropriation. The gigantism, automaticity, capacity, multiple attack points, and amorality of modern technology congeal into a physical machine for mass slaughter and large-scale habitat destruction. The authority and longevity of institutional frameworks entrench human-nonhuman unequal power structures. Thus does the elective affinity of linguistic constructs, technologies, and institutions sanction a stranglehold over the living ocean—a stranglehold that, having been validated by numerous social structures, appears as a normal state of affairs to the proverbial man on the street.

The extermination technologies that do what is still quaintly called fishing include industrially equipped vessels working together as fleets that often stay at sea for months at a time; on-ship machinery for processing and storing amounts of fish so gigantic that one such super-trawler can "feed 18 million people a good supper";[21] a longline industry that sets roughly two billion hooks per year, killing beings indiscriminately; bottom- and mid-water trawlers that can span the width of a football field and the height of a five-story building, which also kill beings indiscriminately and additionally demolish marine habitats; purse seine nets that can be more than two thousand meters in length and two hundred meters in depth, and are capable of ripping out entire schools of fish; aircraft surveillance as well as bottom-imaging technology for tracking down the fish wherever they may be; and detailed seabed maps constructed by corporations and sold to the fishing industry.

The role of modern technology in devastating marine life has been duly noted by all observers of the state of the ocean. Because of relentless improvements in technology over the last century, "fishing fleets have become powerful enough to overexploit essentially all stocks [sic] in the world, anywhere, any time of the year, thereby removing the last available 'natural protection' afforded by depth or distance from the shore."[22] Marine life can find no refuge from this assault, for which *war* is the metaphor even scientists are using. "You could declare war on the world with the technology that's on a tuna boat," states marine biologist Jeremy Jackson.[23] Fisheries scientist Daniel Pauly also warns that "we are fighting a war against fish and we are winning."[24]

A shared human-possessive language makes mass extermination conventional, and technology carries out the task efficiently. What is still needed are the institutional structures that dictate who owns what, legislate what is taken, and ensure the subsequent distribution of the commodities. I turn to peruse some key institutional structures at work.

For starters, 97 percent of the seas are legally fishable. This arrangement harks back to a seventeenth-century treatise called *Mare Liberum*, or *Freedom of the Seas*. This doctrine affirmed the right of all nations to use the ocean as a commonwealth for travel, fishing, and trade.[25] Today, *Mare Liberum* is often portrayed as superseded by the contemporary agreement of UNCLOS (the United Nations Convention on the Law of the Seas). Yet the legacy of *Mare Liberum* is omnipresent, even as it has been modified to establish that certain portions of the seas are nation owned. These "economic exclusive zones" legislate national ownership of two hundred nautical miles from the coast (or further if a continental shelf reaches beyond two hundred miles from a nation's coast). So it is that some of the most fabulous places on Earth— its life-rich coastal seas and continental shelves—have been renamed "economic zones." Unsurprisingly, worldwide coastal seas are critically endangered, and continental shelves are endangered.[26]

Other highlights of institutional arrangements include billions of taxpayer dollars in subsidies to industrial fishing; an international statistics-gathering system that routinely underestimates the global fish catch;[27] legal loopholes for unregulated fishing, such as fishing vessels purchasing "flags of convenience," which are flags of nations with no fishing regulations; nonenforcement, or lax enforcement, of laws against illegal fishing and overfishing; laws that allow developed nations to purchase fishing rights from developing nations, thus devastating the waters where local fishermen were earning a living; regulatory frameworks wherein scientists advise sustainable quotas for fishing, which are jacked up by 50 percent or more by politicians, and which are subsequently "usually disregarded by rapacious fishermen";[28] a free-for-all fishing spree in the high seas; an expanding grid of infrastructures—roadway systems, seaports, railways, canals, and airports to bring fish to more and more outlets; and inaction on the part of governments and the international community in establishing the needed networks of strictly protected marine areas around the globe.[29]

The elective affinity of linguistic ownership, mega-technological gear, and institutional frameworks associated with industrial fishing has transformed the entire ocean into a resource domain. The sway of that elective affinity is such that it has produced a totalitarian regime over the marine realm, while effectively also co-opting the public's consent to that regime. This arrangement has resulted in the commodification of the ocean for mass consumption to serve the freedom (actual or promised) to consume fish affordably, in any quantity, at any time, and in a variety of forms: fresh, frozen, processed, precooked, and at fast-food outlets. Fish is today's most traded "commodity."[30] A quarter of the caught fish become feed for confined domestic animals (like pigs and chickens), industrial aquaculture operations, and pets. One way and another, industrial fishing supports the provisioning of cheap food for increasingly more people.

All the taking has emptied the seas of the livelihood of nonhuman animals. Vacuuming the herring, menhaden, anchovies, sardines, and other small fish takes its toll on the seabirds, sea lions, penguins, and whales, among others. Whales are washing up dead of starvation, and on California's seacoast, so are seals and their pups.[31] The Arctic terns, as well as puffins, razorbills, and other seabirds, are taking nosedives in their numbers. The terns, not finding enough herring, try to feed their chicks butterfish, but the latter are too big for the chicks, who die surrounded by food they cannot swallow.[32] With the krill (for now) a "sustainable fishery," how can whales find enough sustenance to make a comeback? Dolphins starve to death as "the sardines and anchovies they've eaten for ages are now being caught to feed caged farmed salmon and tuna."[33] From an exuberance of life, the ocean has been turned into an unsafe zone for its residents. All of this death and suffering seem an acceptable price for the human freedom to consume unlimited amounts of "seafood."

For nonhumans the regime that has been imposed on the ocean is one of terror and untimely death. This is evident in how fish and other sea creatures ("fisheries") have been massively exterminated ("unsustainably harvested"), by means of a military operation (industrial fishing), underwritten by the institutionalization of the seas as human property—either as a free-for-all expanse or as national districts. The arrangement secured by interlocking social structures is founded on abolishing the freedom of living beings—including their freedom to move, to feed themselves, to experience flourishing, to enjoy lives un-hampered by unnecessary fear and exploitation, to live their natural life spans, to cocreate their watery worlds, and even to remain in exis-tence. *Mare Liberum* has brought about the eradication of their auton-omy, and made their lives precarious, their world toxic, their suffering acute, and their homes profoundly impoverished.

Infrastructure Expansionism and the Plague of Road Building

One way to simplify and at the same time cast a broad net over civi-lization's relation to the land—in the ostensible service of enhancing human freedoms—is to consider the role of infrastructure in our time.

Infrastructure is the necessary factor for all development: it is in-dispensable for moving raw materials (like fossil fuels and rare earths); it is required for connecting more and more people (via internet and cellular phones); and it is needed for human mobility and the flow of global trade. Through infrastructural spread of urban, commercial, and rural build-out, mining operations, highways and roads, dams, elec-tricity grids and power lines, cellular towers, and other developments, the land (as well as the seas) is becoming increasingly occupied. As the human freedoms to move, trade, and communicate are expanded via such development, so do the freedoms of many nonhumans vanish or shrink: their freedom to move, to live in accordance with their natures, to experience the world on their terms, and to continue existing and evolving.

Formal definitions of infrastructure reveal that its fundamental mo-dality is movement. One expert compares infrastructures to "arteries and veins attaching society to the essential commodities and services required to uphold or improve standards of living."[34] Infrastructure has also been defined as public works that "facilitate the transport of peo-ple and goods, provision water for drinking and a variety of other uses, provision energy where it is needed, and transmit information within and between communities."[35] Yet another expert states that infrastruc-ture represents "assets that are involved in the movement of goods, people, water, and energy."[36] At the material level, infrastructure is all about movement—long-distance movement of people, raw materials, commodities, electricity, information, imagery, services, and money. At the symbolic level, the expanding mobility of everything is valo-rized as an asset bringing a widening breadth of choices and actions to people. Indeed, much of what development pragmatically refers to today is building up modern infrastructures in the developing world or anywhere deemed as having an "infrastructure gap." Filling that infrastructure gap is groundwork for bringing the valued freedoms to all.

In a world of 7.5 billion people and growing—people mostly aspiring or cajoled into aspiring to modern freedoms—the march of industrial infrastructure is leading toward the human landscaping the world. I refer here to constructions and networks largely taken for granted as desirable or inevitable: the burgeoning of roads; highways; parking lots; canals; dams; irrigation schemes; power lines; communication towers; airports and seaports; mining operations; oil, natural gas, and coal slurry pipelines; industrial-scale wind and solar farms; food and livestock storage facilities; and conventional or high-speed railways. These form a constellation of interlocking works, supported by an alliance of the public and private sectors, that link humanity at material and cybernetic levels. They are the technological bedrock of the human enterprise's expansionism, the thickening grid upon which civilization plugs itself into the entire biosphere. Gridding the world with an industrial infrastructure that functions as the conveyor belt of all kinds of movement for people, by its very character, foils or eradicates nonhuman freedoms and continues to extinguish what is left of wilderness. Space allows me only partial discussion of infrastructural sprawl: I focus on roads and the general direction infrastructure build-out is heading.

In a review article on the impacts of roads on the natural world, conservation biologist Reed Noss describes roads, from primitive logging roads to four-lane highways, as a "massive tragedy."[37] Roadkill is an obvious and generally deemed acceptable price. The global numbers are likely beyond calculation, but a widely cited ballpark estimate is that in the United States alone one million animals die on roads every day.[38] In his insightful critique of the structural violence of roadkill, sociologist Dennis Soron points out that the ingrained cultural association of the automobile with human freedom militates against seeing it as "a leading agent of violence," thereby sponsoring "the banality of roadkill." "The stark visibility of broken animal bodies on roadways," he writes, "has generated mostly fatalism and disengagement, becoming as naturalized a part of contemporary landscapes as roads and automobile traffic themselves."[39] Thus, the underside of the image and experience of the automobile as herald of freedom is the rendering of graphically violent death as "unseen."

For animals who are attracted to roads for foraging reasons, or to warm themselves, roads are a death trap. For endangered animals (like the Florida panther and desert tortoise, for example), road deaths contribute another significant pressure on their survival. Roads also fragment, and of course replace, natural habitat. For animals who are road averse, roads restrict their available territory and range. For such animals roads are barriers that they refuse to cross, so that one or more roads divide and insulate their populations. Additionally, roads contribute to pollution, facilitate invasion by nonnative species, and (where there are waterways) increase erosion, sedimentation, and pollution.[40]

Yet by far the greatest impact of roads is that they are superconductors for people with guns, traps, ploughs, chainsaws, drills, and bulldozers. The most significant upshot of roads is that they are harbingers of agricultural, extractive, and residential development, ushering in the multipronged demolition of wild nature. Biologist Adrian Forsyth narrates how road-initiated demolition can happen in the blink of an eye: "I can remember," he states, "going to western Ecuador in the early 70's and being at the base of the Andes and looking at the foothills covered with ridge after ridge after ridge of dark forest receding into the distance and just sort of blithely assuming that that would be there forever. And then a road was built from the highlands to the lowlands and 20 years later there were half a million people living there and not a shred of forest as far as the eye could see."[41] Regarding the condition of forests worldwide, he adds: "The forest is pretty chewed up wherever there's a road."

In the United States, roads in the national forests (the biggest timber providers) add up to four hundred thousand miles.[42] The American interstate highway system, including its exits, smaller highways, and roads, has sliced up most of the continent. Regarding impact on the Amazon, a *Mongabay* article informs that "after the construction of the Trans-Amazonian Highway, Brazilian deforestation accelerated to levels never before seen and vast swaths of forest were cleared for subsistence farmers and cattle-ranching schemes."[43] Ninety-five percent of all destruction of the Amazon rain forest occurs within five kilometers of a road.[44] The construction of logging roads is prying open the Congo rain forest to subsistence and export agriculture, industrial logging, mining, and poaching. "Logging roads," reports journalist Rhett Butler, "have opened up vast areas of the Congo to commercial hunting, leading to a poaching epidemic in some areas and a more than 60 percent drop in the region's forest elephant population in less than a decade."[45] Moreover, as the Congo basin becomes more accessible by roads, the perilous expansion of industrial agriculture, such as oil palm, soybean, and sugar plantations, becomes more real. Meanwhile, the plan of a four-hundred-kilometer road network in northern Sumatra would open up to poachers, and inevitably to development, the last refuge for its tigers, orangutans, elephants, and rhinos.[46]

The ecological effects of roads—killing, habitat destruction, wilderness extirpation, and extinction—constitute their "externalities," meaning the impact impinged on parties who do not choose to incur the impact. Wild nonhumans, and wilderness overall, are the parties who do not choose to incur the destructiveness of roads. They do not count. But as far as the mainstream human domain is concerned, road infrastructure is all about "if you build it, they will come": people tend to take advantage of the privileges roads provide, such as greater mobility; wider habitat range; and access to farming, hunting, fishing, mining, recreation, and all manner of development.

Like other externalities of the human impact on the biosphere, the ecological harms of roads go ignored. The only way to avoid the externalities of roads is to not build them, as well as to undo many roads already built.[47] The global trend, however, is in the opposite direction. Since 1970, worldwide roads have grown fourfold. With respect to the future, according to a 2013 Report of the International Energy Agency, "global road traffic activity is expected to more than double by 2050. To accommodate this growth, global road infrastructure is expected to increase by roughly 60 percent above 2010 levels."[48] Notably, these growth figures pertain to paved roads. There exist little to no official data for trends regarding unpaved roads, which are the frontline of invading the biosphere's remaining terrestrial wilderness—and which eventually often get paved.[49]

No matter their ecological costs, from a conventional standpoint roads are among the most valued infrastructure; their numbers and condition in any given nation speak to that nation's perceived level of development. Development banks typically call roads "the backbone of the economy." For the International Road Federation—established in 1948 with the motto "Better Roads for Better Living"—roads remain the number 1 mobility vector that also generate business contacts. "Governments are aware that an underdeveloped road network is likely to be associated with sub-optimal economic performance and quality of life," according to the World Bank Group.[50] The words "integration" and "prosperity" are the most commonly rehearsed to describe the enhancement of conditions associated with building up transportation networks.

Circling back to Noss's paper, the ecological effects of roads vary, but, in his words, "virtually all are bad, and the net effect of all roads is nothing short of catastrophic."[51] Their catastrophic effect, however, is trumped by the augmented choices and actions for people that road building represents. Roads enable valued personal mobility and secure the connectivity demanded by growing international trade and travel. Thus, worldwide there is no end in sight to road building. And more roads become imperative as the number of vehicles swell with economic development.

Even so, as biologist William Laurance notes, "roads are just part of the problem. Everywhere one looks, new infrastructure is proliferating in the world's last wild places, often provoking serious environmental harm."[52] Indeed, as of 2014 the intent to stimulate the global economy has inspired an initiative to pour trillions of dollars into infrastructure development. This is being spearheaded by the leadership of the developed world, backed by a multitude of development banks on all continents, embraced by emerging economies, and eagerly acquiesced to by developing nations. While the expansion of industrial infrastructure is nothing new, "the new plan's scale" and "breathtaking speed" are unprecedented.[53] "In each geographic region—as well as globally—plans are underway to create, strengthen, and expand infrastructure for enhanced trade and integration. These plans are interdependent since expanded trade depends upon infrastructure to mine or move raw materials, manufactured goods, and services."[54] Asia alone plans to pour $26 trillion into new infrastructure—including China's already launched Belt and Road Initiative to grease trade between East Asia and Europe—by 2030.[55] These infrastructure expansions—intended to disseminate the freedoms associated with a modern lifestyle—harbor as their mere side effect the all-out humanization of the planet. While the natural world can present physical challenges to infrastructure development, ethically and existentially it largely appears as empty space.

Limitless expansionism is a logical upshot of the human-supremacist worldview, which recognizes little of inherent ontological substance or irreplaceable value to bar its sprawl into the nonhuman domain. The escalation of mega-infrastructure in our time is a signal manifestation of this worldview. Contrary to the way it is pitched, infrastructural expansionism is not a pragmatic program, for infrastructure needs could be served with small-scale and decentralized works. What's more, pressures for more infrastructural development could be alleviated by reducing global trade (especially of throwaway and unnecessary commodities), by supporting local and regional economies, and by prioritizing human rights–enhancing policies to stabilize and reduce the global population (discussed in chapter 8). Additionally, large swaths of natural areas could be placed off limits to infrastructural expansion to protect the livelihood of nonhumans and the integrity of wild habitats.

Rather than a pragmatic imperative, the contemporary spread of infrastructure—especially, the campaign for "more mega-highway projects, more centralized electric power plants and electric grids, more mega-dams and gigantic irrigation schemes for water transfers"[56]—defers to the ideological directive of human supremacy for a number of reasons. One, the spread of infrastructures vaunts the lack of restraint over Earth's landscapes and seascapes, thereby, even in this time of crises, reasserting the absolute sovereignty of human agency within the biosphere. Two, infrastructural expansion also works to display and inflate human (and nationalistic) pride via the conspicuousness of gigantic works, such as gargantuan dams and other mega-technological constructions, requiring impressive engineering aptitude and an immense orchestration of labor. And last but not least, infrastructural build-out flaunts human ownership of the planet by thrusting humanized landscapes into the perceptual and experiential foreground, while resetting natural scenery as the backdrop of human works. Mega-technological impositions on landscapes and seascapes thus function as "massive monuments" of and to human power.[57] They are the pyramids and cathedrals of our time, "psychologically inflated," in the words of Lewis Mumford, "by a similar myth of unqualified power, boasting through their science of their increasing omnipotence, moved by obsessions and compulsions no less irrational than those of earlier absolute systems: particularly the notion that the system itself must be expanded, at whatever eventual cost to life."[58]

In sum, the social structures of linguistic renderings, technological means, and economic, political, and legal institutions fit seamlessly together: as a potent unified force, it succeeds in entrenching the regime of nature's domination, in instilling an unmistakable message about its surety, and in locking down that surety by means of producing and reproducing nature's domination as consensual reality.

The subjugation of the nonhuman domain also draws powerfully on the social body's collective amnesia of the complexity, abundance, and variety of life that constitutes the primordial essence of the biosphere. As discussed earlier, this amnesia is socioculturally orchestrated by means of the dominant culture's dearth of traditions to record and celebrate the participatory presence of nonhuman ecologies and neighbors. The orchestration of amnesia is a crucial feature of the human-supremacist silencing of the natural world, for human oblivion about the fate of the natural world conspires to exclude its acknowledgment as an intrinsic domain.

Yet even when the biosphere's original richness is remembered, mechanisms exist to defuse its subversive potential to incite and inspire a human desire to inhabit that richness. A signal giveaway of the consensual reality of nature's domination reaffirming its authority is excusing the capitulation to nature's domination as "realism." For example, authors Peter Kareiva and Michelle Marvier exhort that "some realism is in order" in conceding that species like "wolves and grizzly bears" (and, by implication, American bison, Tibetan antelopes, sea turtles, seabirds, cod, sharks, whales, and so on) "will never be as abundant and widespread as they once were."[59] (Never mind that "never" is too long a time for such a wager—"never" has a pointed effect to produce.) In such exhortations, the human refashioning of the world is portrayed as a de facto baseline or fait accompli not to be questioned by reasonable people. While this kind of urged concession is an effect of power—the power of the domination regime speaking—such concessions do not acknowledge that they are genuflecting before power's ironclad sway. Instead, resigning ourselves to the subjugation of the natural world is represented as deferring to reality—"reality" being *the* existential category that rational people do not go around objecting to.

The status quo of Earth's domination demands of us to surrender to the imposed design upon, and downgraded version of, the biosphere it produces. For those who have "forgotten" nature's primal wealth (the majority), surrender is even easier; the subjugated biosphere and the real world have become thoroughly fused. But should that primal wealth of life in the seas and on the land be recollected, reasonable people must be realistic and let it go—whether the "it" be memory, grief, desire, or mutiny against consensus reality.

Mobility Denied

Movement is one of the fundamental freedoms taken away from many nonhumans. This follows ineluctably from the fact that only humans may define and control geographical space. Even as geographical space is carved up and repurposed to enhance the movement of stuff and people, so, as a direct corollary, is the mobility of nonhumans refused. The large-scale movements and migrations of big carnivores and herbivores are especially unwelcome and vastly curtailed. As critical geographer David Lulka puts it, "mobile individuals have been viewed as threats to the health, integrity, and efficiency of existing social systems."[60] Today, wide roving and migratory beings are increasingly finding their journeys foiled and their destination places filled with dangers and challenges.

Animal voyages are phenomena of abundance. They are epic and mind-boggling in the skill, effort, and intent they exhibit. Billions of songbirds, seabirds, waterfowl, and raptors make seasonal journeys. So do certain butterflies and dragonflies, traversing oceanic and continental distances. Whales, sea turtles, sharks, tuna, and other sea creatures travel the ocean in different life stages and seasons. Caribou, pronghorn antelope, elk, and mule deer still undertake migrations on the North American continent, though most of their ancient routes have been erased. The same applies to the chiru and wild yaks of the Tibetan plateau and the saiga of Mongolia. Some snakes and salamanders also make journeys. Multitudes of fish have meandered between seas and land, weaving together ecologies and distributing nutrient flows. Their runs were once "more fish than water."[61] For thousands of years, migratory fish fed places with their feces, sperm, eggs, offspring, and living and decaying bodies, but their "once great spawning runs have slowed to a trickle."[62]

"Around the world," note ecologists David Wilcove and Martin Wikelski, "many of the most spectacular migrations have either disappeared due to human activities or are in steep decline."[63] Moreover, the rate at which migratory species are declining has accelerated in recent years.[64] While the constellation of specific causes of declines vary by species and place, underlying them all is humanity's expansionism to serve a plush buffet of choices and actions for people. The unlimited human freedoms pursued efface the freedoms of many nonhumans to move, express their identities, complete their life cycles, take full pleasure in being alive, and craft landscapes and seascapes with their abundances and peregrinations.

Whether or not the disappearance of migratory phenomena will lead species to extinction (which can certainly occur), what is endangered is their abundances and ways of life. Animal migrations are primarily about two things: teeming numbers and freedom of movement. Neither of these dimensions is permitted coexistence within the regime that humanity has imposed in the biosphere.

In their dispersals and migrations, animals face threats en route, in the places they stop to rest and refuel, and in their seasonal homes. They encounter the conversion and fragmentation of their homelands, a raft of physical barriers, targeted killing, and increasingly unreliable conditions due to a rapidly changing climate. Habitat is converted into agriculture, mining operations, or coastal resorts and settlements, among other developments. Obstacles include dams, highways, fences, cellular towers, power lines, and fishing gear. Migration routes and stopovers sometimes provide opportunity for people to kill the animals. (Gray whales were almost driven to extinction in their breeding and calving grounds of Baja California. Hunters kill songbirds by the millions in Cyprus and Italy, as the birds travel between their wintering grounds in Africa and summering places in Europe.) Climate change is anticipated to eat up critical habitat such as many remaining nest beaches of sea turtles; to discombobulate phenological timing, such as when animals and their prey species get out of synch with the scrambling of the seasons; and to alter the hydrological conditions for creatures sensitive to those conditions, like amphibians.

The pronghorn antelopes of North America have lost nearly 80 percent of their migratory routes to development. Those who remain persist in making their three-hundred-mile roundtrip ancient journey, forced to circumvent highways, housing settlements, over one hundred miles of farming fences, and more recently landscapes subjected to fracking.[65] The saiga antelopes have declined by 95 percent. Their migratory routes are stymied by further increases in human numbers and affluence, as well as by development (especially mining) in their ecoregion.[66]

Meanwhile, the populations of many migratory songbirds crossing the Gulf of Mexico are falling.[67] Among them cerulean warblers "are declining at an alarming rate" having dropped 80 percent in forty years.[68] This reflects habitat loss at both ends of their migratory routes. "Mountaintop removal mining," Wilcove explains, "an extraordinarily destructive practice in which tops of mountains are scraped away to expose coal seams, has already destroyed hundreds of thousands of acres of breeding habitat in the Appalachians. . . . Much of the warbler's wintering habitat [in Peru and Venezuela] has been converted to cattle pastures, coffee and coca plantations, and other agricultural uses."[69] Physical barriers also take their toll. Even the common Swainson's thrush (listed by the International Union for Conservation of Nature as "least concern") experienced a 38 percent decline between 1966 and 2014. According to the Cornell Lab of Ornithology, "during spring and fall migration significant numbers of Swainson thrushes die from collisions with windows, radio and cell-phone towers, and tall buildings."[70]

European-African songbirds are also in distress. A variety of them such as robins, swallows, and warblers summer in Europe and winter in Africa, traversing thousands of miles in between. The spread of agriculture, grazing, and desertification in Africa is destroying their places of livelihood and food supplies, while en route they are shot for "bushmeat" or sport.[71] The pied flycatcher, a small bird breeding in the Netherlands and wintering in Africa, has precipitously declined: caterpillars in Europe are emerging earlier and earlier, so that the peak of the caterpillar population has gotten out of step with the pied flycatchers provisioning their nestlings.[72] This is one among many similar cases of ecological relations beginning to come undone from climate change. Songbirds of the Americas, for example, are being hurt by the same emerging mismatch of life cycles: the longer their journeys, the greater their declines.

"If you are a resident of the East Coast of the United States or of Western Europe," asks biologist John Waldman, "when did you last attend a shad bake, eat an eel, or watch salmon vault a waterfall?" "Numerous measures show," he adds, "that two-dozen migratory fishes of both shores of the North Atlantic have seen profound reduction."[73] In the last thirty years, Canada's wild Atlantic salmon population has dropped 75 percent.[74] In Finland, wild Atlantic salmon have disappeared from 90 percent of the rivers where they spawned historically, while in France, they have vanished from a third of their historic rivers and are endangered in the remaining two-thirds.[75] Salmon species of the Pacific Northwest are less than 10 percent of their numbers prior to European settlement; dams, overfishing, diversion of water for agriculture, and streamside farm animal grazing are the causes of their extermination.[76]

Humpback whales migrate 2,500–4,000 miles between Hawaii and Alaska. They are threatened by fishing nets and lines, as well as collisions with cargo vessels and cruising ships.[77] The remaining North Atlantic right whales also experience increased mortality by collisions with ships and entanglement in fishing gear. They migrate along one of the busiest coastal transportation routes, yet there has been "no serious effort to reroute commercial traffic away from the migratory route of whales," nor to reduce the fishing gear in the region.[78]

Writing about Africa, the last continent where big migrations persist, Wilcove anticipates the compounded consequences of growing human numbers, increasing prosperity, expanding infrastructures, and the takeover of landscapes, especially for agriculture, grazing, and mining. "The end result," Wilcove writes, "will be an Africa that looks increasingly like the rest of the world, largely devoid of spectacular mammal migrations, its large mammals restricted to a small number of major parks, the parks themselves encircled by fences, what little is left of the wilderness incarcerated for its own safety."[79] Indeed, Africa is heading precisely in that direction through a combination of economic globalization ventures and explosive population growth.[80] Already, its remaining lions are only safe in protected areas.[81]

Humanity has attained control of geographical space by obliterating wilderness—the untamed biogeography that has its own will, its own infinity, its own destiny, and its own imagination—into fragmented plots, islands of isolated nature, and souvenirs of the bygone past. An integral dimension of the physical and conceptual erasure of the wild—in its original state of unconquered, expansive spatial being—is the control of geographical space sustained by the eradication and/or management of the world's nonhuman mobiles: animals who live and move in packs, herds, flocks, and schools.

Some animal migrations have already been completely eradicated. Wilcove recounts that "South Africa once hosted a mammal migration that rivaled what we still see in the Serengeti," describing it as "unquestionably one of the continent's greatest wildlife spectacles."[82] This migration included quagga zebras (a subspecies hunted to extinction), black wildebeest (persisting today, much like American bison, in a semidomesticated state in ranches and reserves), and most especially the springbok antelope who once numbered in the hundreds of thousands. Colonial hunters extirpated the regional predator, the Cape lion, in the nineteenth century. South Africa's migration is gone from both world and memory. Fencing the land, livestock grazing, and hunting by European settlers destroyed it. Furthermore, industrial agriculture development is right now endangering Africa's second largest mammal migration in the Ethiopian grasslands of Gambella on the border with South Sudan.[83] Nor does worldwide fame guarantee that the Serengeti migration is safe from development—with plans to build a mining-facilitating road through the animals' migration route in the offing.[84]

The United States annihilated two of the greatest migratory phenomena on Earth: those of the American bison and passenger pigeon. The passenger pigeon was persecuted to extinction, while the American bison, as a wild and nomadic animal, barely survives. The American bison narrowly escaped extinction during the late nineteenth century, when in a span of a few decades their ranks were decimated from tens of millions down to a few hundred.[85] The bison's European relative, the auroch, was not as "fortunate." The species was exterminated from the face of the planet by the ancestors of the European colonizers.

It is common knowledge that the American bison's extermination served to subdue the Plains Indians by destroying their main source of livelihood. True enough, but more to the point *both* the bison and Indian near genocides were required in order to take over the grasslands and plow them under.[86] Industrial-scale agriculture cannot coexist with twenty, thirty, or more million thundering, ever-on-the-move beasts, each weighing between seven hundred and two thousand pounds, and entraining an enormous ecological cohort from prairie dogs and pronghorn antelopes to grizzlies and wolves. Today, bison numbers have slowly risen to about five hundred thousand, but the majority live in ranches, and many have been crossbred with cattle; they have been all but domesticated, behaviorally and genetically.[87] With the fate of the bison, among other large mobile animals, we witness that "what is gone is not the species but *the phenomenon* of the species."[88]

The only free-roaming bison herd in the United States, in Yellowstone National Park, is not exactly free to roam. Bison are nomadic by nature and often venture beyond the park boundaries. Such behavior is not tolerated, and a primary management objective is to haze bison back within park boundaries. When nonlethal management methods fail, the method of dealing with errant bison is to shoot them.[89] When bison numbers grow beyond what is deemed sustainable for their delimited territory, the method of dealing with their increase is to cull them. As Wilcove observes, "it would appear that even in Big Sky country, there is no longer room for a remnant of the American bison's great migration."[90]

By no means is this just the story of the American bison. For example, the bontebok, South Africa's rarest antelope, was once migratory and abundant. Its existence is today confined to a tiny park, called the Bontebok National Park. When bontebok numbers press against the park's boundaries, the animals are culled. Writing about his visit there, Wilcove reports, "the place seemed more like a zoo than a park and the Bontebok more like cattle than wild animals. They face no enemies, their numbers are regulated by vigilant wildlife biologists, and their movement is constrained by fencing."[91] Indeed, wherever large animals have shaped geographical space by means of the journeys of their teeming numbers, both their abundance and their mobility have (with rare exceptions) been extinguished.

The human regime deprives bison, bontebok, and so many other wild on-the-move earthlings, the freedom, as Lulka aptly puts it, "to inhabit the landscape on their own terms."[92] It also denies them the freedom to contribute to shaping landscapes and seascapes by means of constructing niches for themselves and others; the freedom to inter-act according to their ecological and evolutionary heritage; and the freedom to determine their individual identities and experiences in the world. Yet depriving nonhumans of this gamut of experience and existence is invisible, or a nonproblem, for much of the human collective, because the human-supremacist worldview has barred the realities of freedom, agency, relationship, identity, and self-determination from having meaningful applications in nonhuman life. This effect of supremacy is expedient, for beings cannot be denied what they have been decreed not to have.

Lulka argues that "as spaces designed to resolve human-nonhuman conflicts," protected areas "are a backhanded compliment in their insistence that nonhumans refrain from contesting the social hierarchies imprinted in the geographical landscapes." "Nothing is lost," he adds, "but autonomy."[93] A paradox thus confronts us. On the one hand, protected areas (like national parks and wilderness reserves) are landscapes of confinement intended to restrict wild animal numbers and movement, delimit autonomous ecologies, and contain wild processes within patrolled borders. On the other hand, in the words of conservationist Tom Butler, "protected areas and wildlife protection laws are the key tools for combating human-caused extinctions of our fellow members in the community of life."[94] Hence the paradox: protected areas work to imprison wild expressions of the natural world, yet simultaneously they are indispensable and good for biodiversity.

In a pathbreaking paper titled "The Incarceration of Wildness: Wilderness Areas as Prisons," environmental philosopher Thomas Birch analyses this paradox at length.[95] Birch uses the word "imperium" to describe human control of geographical space, including control of protected areas. He argues that protected areas are the imperium's best gesture toward the more-than-human world, for they represent an extension of the liberal tradition of self-determination to some fraction of that world. At the same time, Birch points out, "designated wilderness areas become prisons in which the imperium incarcerates unassimilable wildness in order to complete itself, to finalize its reign."[96] Protected areas comprise territories grudgingly conceded to the natural world, just as reservations have been to defeated native peoples. The great Sioux chief Black Elk noted the similarity, writing that when "the Wasichus [white men] came they made little islands for us and other little islands for the four-leggeds, and always these islands are becoming smaller, for around them surges the gnawing flood of the Wasichus."[97]

Such "little islands" stand as emblems of control over the wild. They conveniently also perform as tokens of tolerance and beneficence, even as they are contoured to fit within the modern order: protected areas augment the variety of consumer options, enlarging modernity's proffered freedoms as places serviceable for recreation, stress reduction, spa services, or providers of peak experiences. The paradox here is that while some wilderness is circumscribed to remain relatively untouched, at the same time it is assigned a delimited place and assimilated into the buffet of riches that the modern world has to offer. The way the latter framing succeeds is subtle: there exists a widespread perception that a human-dominated world order, peppered here and there by some protected areas of land and sea, is a benign blueprint for the biosphere's design in perpetuity. Birch captures the paradox of incarcerated wilderness as follows: "Wilderness reservations serve as a crucial counter-friction to the machine of total domination. . . . But insofar as wilderness reservations, as they are so often (mis)understood, only serve the completion of the imperium, they are not justifiable."[98]

At this historical juncture, strictly protected areas inhabit a nebulous zone. They will end up as little more than glorified zoological parks, if their ontology solidifies according to the dominant meaning: confined wild nature curated as exhibits of deep natural history, and useful as scientific research centers, experience and vacation providers, resource reserves, sustainable-use manors, and ecosystem-service suppliers.

This need not be, however, how we think of protected areas nor their fate. More profoundly, as Birch urged, they are "holes and cracks . . . 'free spaces' or 'liberated zones,' in the fabric of domination and self-deception that fuels and shapes our mainstream contemporary culture."[99] Protected areas are not ultimately places for managing and policing wild nature—keeping it tidy with a stream of herbicides, culls, and regulations. They can become domains to found upon and expand outwardly from, spreading wildness and beauty and the freedom of nature's boundless creativity over the biosphere. They foster the possibility of liberation for both nature and humanity, for they hold the promise of Earth restoration.

The Reaches of Freedom

Human expansionism in the service of enhancing human freedoms is incoherently premised on divesting the more-than-human world of its liberty. At its deepest layer, the crisis of life in our time is a crisis of freedom. For nonhumans, it is a crisis of the *condition* of freedom, the freedom to live in accordance with their natures and to experience flourishing. For humanity, it is a crisis of the *idea* of freedom, which degenerates into a specious, shallow, and muddied-by-egoism unrestricted and entitled sprawl.

Extinctions of species, subspecies, populations, places, and nonhuman ways of living are ubiquitous, but many people cannot see them, do not want to see them, or prefer to forget or dismiss their significance. Yet as eco-theologian Thomas Berry submitted, it is a great illusion of our age that human life can be enhanced by diminishing the more-than-human world.[100] In the same vein, it is a great illusion of our time that limitless human freedoms pursued at the price of the constriction, destruction, or enslavement of the larger community of life can bring humanity lasting fulfillment and self-realization, or lead toward a more enlightened civilization.

We must deconstruct human entitlement (masquerading as freedom) while simultaneously opening toward an authentic and expanded understanding of freedom. Lighting up the subjugation of the more-than-human world is only part of the work. Another crucial piece is to envision a human way of life that will enable the realization of a broadly shared ideal of freedom for all inhabitants of the biosphere. We must think deeply about what human freedom looks like within the liberated expanse of a living planet. While this may appear to present a gloomy exercise in human self-restrictions, such would be a superficial assessment. Humanity's willful embrace of limitations harbors the actuation of a more evolved civilization and a high quality of life, because authentic human freedom will never be achieved by trampling over the freedom of all else.

Authentic freedom includes freedom for all life, not only for humans and not only additionally for wild and domestic animals, who are our closest kin. In the words of deep ecologist Arne Naess, "the intuitive concept of 'life' (or 'living being') sometimes includes a river, a landscape, a wilderness, a mountain."[101] To arrive at an enlarged vision and practice of freedom necessitates opening to unfamiliar emanations of the idea, emanations often relegated to the sphere of the merely poetic or fanciful. We must cultivate new ways of speaking, new ways of seeing, and a new imagination of Earth and ourselves within it. "Language changes and imagination is on our side," Jack Turner exhorts. "Perhaps in a thousand years our most sacred objects will be illuminated flora, vast taxonomies of insects, and a repertoire of songs we shall sing to whales."[102]

Wherever we find, through inner or outer censorship, a dismissal of an enlightened vision for human life as unrealistic or romantic, there we are witnessing the supervision of the human imagination, which is allowed expression only insofar as it "transforms reality within the general framework of repression."[103] Thus, for example, the imagination is encouraged free range in designing a technically intricate mega-dam or figuring out increasingly precise methods of genetic engineering. But when the imagination becomes receptive to life-embracing affirmations—such as "No yard! but unfenced Nature reaching up to your very sill" or "Burn the fences and let the forest stand!" or "Let the rivers run free!"[104]—such receptivity suffers disparagement as beyond the bounds of the real world. But that "real world" in the name of which the imagination is censured from moving outside its cage is a biophysically materialized and ideationally diffuse supremacist worldview that has entrenched itself by usurping reality.

Freedom is the highest value in existence because it is the indispensable condition for the realization of the full potential of being and becoming. The pursuit of unlimited human "freedoms" within the biosphere is a mockery of that highest value. Just as the freedoms slavery enabled for slave owners, and for consumers of the cheap products of slave labor, today appear superficial and are no longer respectable, so we are compelled to conclude that modern freedoms—their horizon-expanding, wealth-accruing, and consumer-enhancing appearances notwithstanding—exercised at the expense of the nonhuman domain are entitlements misconstrued as freedoms. Restricting and relinquishing such freedoms is therefore hardly a sacrifice: it is the path toward an inclusive understanding of freedom and multispecies co-flourishing within the biosphere.

Creating an ecological civilization is foremost about honoring freedom in its authentic and expanded meaning. The quest for real freedom can ignite the imagination toward new ways of inhabiting Earth: respecting limitations in terms of how much land and seas humanity occupies and uses; redesigning how landscapes and seascapes are inhabited and used; rethinking how many human beings a liberated planet can support; and reorganizing economic life, activities, and relations. Such aspirations, grounded on the realization of freedom for all, will steer humanity's endeavors away from the impending dystopian world that the human enterprise's never-ending growth is manufacturing.

PART THREE

Scaling Down and Pulling Back

SEVEN

Dystopia at the Doorstep

Those endeavors that produce food and energy need scale and landscape
that are of necessity rural and are of necessity unspeakably destructive.

RICHARD MANNING

A global population of over ten billion people, by the end
of this century, is often presented as though it is as inevitable as the trajectory of an asteroid hurtling through space.
This oft-rehearsed projection signals a widely shared fatalism: the estimated growth is deemed as having too much
inertia behind it, and as being too politically sensitive, to
question. At the same time, the repetition of the projection reinforces the impression that nothing can be done
to change it. The incantation of "ten billion" seems at
work as self-fulfilling prophesy, for by doing nothing, it is
exactly where we are (minimally) headed.[1] So do we collectively hypnotize and propel ourselves in the predicted
direction.

Environmental analysts have divergent responses to
this particular figure. Some are incredulous that such a
number can be approached—let alone sustained—and contend that continued advance in that direction will be catastrophic: some global disaster, or series of climate-related,
food-related, and/or conflict-related disasters around the
world, is bound to derail demographers' projections, and
humanity (after perhaps experiencing a wake-up call) will
stabilize at lower numbers.[2] Other environmental observers, however, are contemplating strategies that might sustain the expected billions. They hope that by raising productivity on lands already under cultivation, and through

efficiency measures and innovations in crop genetics, irrigation systems, pesticide and fertilizer applications, and food waste management, coupled with requisite energy transitions and other developments, the planet might support the coming more than ten billion.[3] There is reason to wager, they maintain, that humanity might succeed at the task, since people are resourceful and determined in the face of adversity.

Thus, where some see disaster or even civilizational collapse on the horizon, others contend that with another techno-managerial leap of progress humanity might avert harsh penalties to continued population and economic growth. Despite divergence in outlook, all analysts agree that even as global numbers and affluence keep climbing, grueling challenges lie ahead, each immense in its own right but daunting in their unpredictable synergies: biodiversity destruction, climate change, freshwater depletions, ceilings on agricultural productivity, all manner of pollution, topsoil loss, and ocean acidification to mention some prominent examples.

Rather than taking sides on the classic debate between the forecasters of impending doom versus the optimists about "feeding the world" (or about a high-tech, consumer-rich global civilization around the corner), I offer a different lens on the future's story. The key issue is not whether it is (or is not) biophysically possible for ten or more billion people to eat industrial food, commune with iPhones, share Amazon wish lists and Spotify playlists, and enjoy a comfortable standard living and global vacation destinations. The point to concentrate on, instead, is that such a world of billions of consumers will only be possible by turning a life-abundant planet into a human resource base, built on extinctions, gridded with industrial infrastructures, webbed together by networks of high-traffic global trade and travel, in which remnants of natural areas—simulacra or residues of wilderness—are zoned for ecological services and ecotourism and managed 24/7.

A geopolitical status quo of over ten billion consumers will require mega-technological support to be sustained (*if* it can be): offshore dikes, sea gates, and other engineering projects to ward off rising seas and frequent storms; more mega-dams for agriculture and energy;[4] the global spread of fracking and deep-sea drilling to serve large-scale energy production; gigantic off-coast and mountaintop wind farms to mass-produce energy; desalination plants, with needed transport infrastructures, to offset water shortages; scaling up of industrial aquaculture to make up for the deficit of ocean fish; genetic modification of crops and animals to adapt to climatic conditions and consumer niches; cultivating so-called marginal lands to grow food or grasses and other plants as biofuels for a gargantuan global car fleet; possible climate engineering at global or regional scales to preempt and/or manage menacing weather patterns; the proliferation of concentrated animal feeding operations (CAFOs) to manufacture mass-produced meat and other animal products; and other large-scale engineering projects, mega-technological systems, and industrial ventures.

The dystopia at the doorstep is an emerging world that—as the Invisible Committee so perfectly nails it—"holds itself together only through the infinite management of its own collapse."[5] Moreover, as we are already seeing, the constant exposure in this world to all manner of man-made and civilization-driven disasters, hazards, risks, and dangers will be met with a demand of "resilience,"[6] wherein people are enjoined to brace themselves and take it, rather than to aspire to build lives free of unnecessary insecurity and suffering.

In such a coming world, private corporate and state capitalisms are bound to entrench their reign, for the required technological gigantism, along with the escalation of mass consumption, will collude to make corporations indispensable. Corporate expertise, corporate research and development, corporate prospecting (for energy and mineral sources), and corporate products will be needed to keep sundry techno-fixes rolling in, as well as various forms of "bread and circuses" (chicken nuggets and Netflix) for the masses. Corporations will generate even more enormous revenues than they do today, via government subsidies and via catering their commodities to billions of people. As author Raj Patel puts it, "corporations may be the creatures of the modern market economy, but in order for profits to flow, they need to conscript consumers to the market."[7] Indeed, there is a snug fit between consumer-population size and private-sector opulence, conspicuous in the correlation between today's richest industrial sectors and their burgeoning middle-class clientele.

The continued growth of the global middle class—the mainstream's basic strategy for creating greater equity among people, or social justice à la the *Economist*—will serve to make the financial and corporate giants of banks, energy, media, entertainment, electronic gadgets, food, agricultural commodities, pharmaceuticals, apparel, and real estate wealthier and more powerful than they already are. Thus, even as the current trend toward curbing extreme poverty continues into the future, the gulf between rich and poor, between the gilded few and the hand-to-mouth many, will persist—precisely because it is the hoi polloi in the middle (and their governments) who are filling the coffers of the (presumed as) privileged "one percent." Additionally, whatever relatively wild places remain will be slated as the real estate and vacation destinations of the most affluent—as they are, after all, already today.

In such a world—whatever it augurs for humanity, which seems sadly bleak—biodiversity will suffer a tremendous blow. Life's richness will not survive a world that is a magnification of the one we live in—with more industrial agriculture, growing livestock numbers, more global trade, more industrial infrastructure, and greater materials and energy consumption (even with efficiency gains)—let alone a world where, in addition, climate-related catastrophes and unpleasant surprises are expected. As discussed earlier, while biodiversity is often equated with numbers of species on Earth, it is far greater than a species inventory. Life is bewildering in its creative expressions, its beauty, strangeness, and unexpectedness, its variety of physical types and kinds of awareness, and its dynamic and interweaving forms of worldmaking. This unfathomable wealth and infinite source of well-being is what humanity is witlessly forfeiting in exchange for heading toward maximizing its population size and consumer spending power.

There exists another choice, a different path forward: The path of halting and contracting human expansionism. Humanity can choose to live on a planet of diverse, abundant, and complex life, instead of haplessly plunging toward a mass-colonized, technologically managed, and corporate-governed planet treated as human property. To live on a planet of life it is necessary to embrace limitations, in order to give the biosphere the unbounded space and freedom it requires to express its ecological and evolutionary arts. For that, we must give the concept of freedom comprehensive scope—pushing its territory beyond the sheath of human exclusivity. In the name of a higher freedom encompassing Earth and all earthlings, human beings can choose to let the greater portion of the world be the magnificence it intrinsically is. Borrowing words from nature writer Julia Whitty, this is the path of intimacy with the planet, taking as our beloved the way things really are and finding our way home.[8]

But the wisdom of limitations—of human and livestock numbers, trade, economies, and places of habitation—is rarely entertained in mainstream thought for what it is: the elegant way home and the surest means for addressing the catastrophes of extinctions, ecosystem losses, rapid climate change, and freshwater and soil depletions. It also happens to be the safest pathway not merely for "feeding the world," but for providing all human beings nutritious, Earth-friendly food. The path of limitations is rarely entertained, because the human-supremacist compulsive expansionism has all but killed the imagination for it. The path of limitations is also sidelined because it is labeled unrealistic and politically inexpedient. But knowledge of the multiple and mounting stresses on the biosphere, along with an understanding of the volatile ways these may compound one another or yield uncontrollable snowballing effects,[9] impels the recognition that drastically scaling down the human enterprise is, in fact, the only realistic approach to the current predicament.

Can the Earth Feed Ten Billion?

In the meantime, even as the available option of limitations is brushed aside, the prevailing question voiced with increasingly urgency is: Can Earth feed ten billion people? By most expert accounts, because of human population growth, alongside the rise of meat and animal product consumption, food production will have to roughly double by 2050 (over 2010 levels) to meet demand, and the big question is: Can it be done? There is an effort under way to figure this out, by experiments in corporate labs, work in research stations, and scrutiny of agricultural databases. And because it is well known that most (and certainly the most fertile) arable lands are already in cultivation, and that the areas where wild creatures live are already hugely constricted, the effort to increase food production—to triple it by century's end—is invariably escorted by the caveat that it must be done without further damage to biodiversity.[10] This approach is known by the oxymoron of "sustainable intensification." "There is a pressing need," in the words of a Royal Society working group, "for the 'sustainable intensification' of global agriculture in which yields are increased without adverse environmental impact and without the cultivation of more land."[11]

Since at least the early 2000s, this ecologically correct sound bite has been activated in environmental writings, journalistic reports, TED talks, and corporate web pages: More crops for food, feed, and fuel must be produced, as well as more meat, dairy, eggs, and fish, by means of careful planning and management, with minimal to no additional ecological impacts. This imperative is to be implemented globally, sometime in the vague future.

The disclaimer of "more food, no additional impact" flies in the face of every present and mounting trend. Sustainable intensification is offered as the needed solution, as if tropical forests are not today giving way to soybean monocultures, cattle ranches, and plantations of oil palm, sugar, tea, and the rest;[12] as if leasing and buying millions of hectares of land in Africa, South America, and elsewhere (by both homegrown and foreign elites) is not already under way in pursuit of food-production expansion;[13] as if marine life does not continue to be vacuumed by industrial fishing; as if rivers are not today so taxed by damming, extraction, and diversion that the crisis of freshwater biodiversity is possibly the gravest extinction site on Earth. It turns out that in the last twenty-five years a tenth of Earth's remaining wilderness was destroyed and "there may be none left within a century if trends continue."[14] Despite all flashing neon signs to the contrary, those at work to figure out if food production can be augmented—to serve a world of billions of consumers in a more tightly knit global economy—always add that it must be done without additional ecological damage. When we encounter such solemn declarations of intent, we'd do well to recall Hamlet's response to the question, "What do you read?" Words, words, words.

The intention to increase food production without more harm to nature, while sincere, is wishful thinking. For even if for a moment we ignore the fact that present-day industrial agriculture, industrial aquaculture, and industrial fishing already constitute a planetwide disaster, simply aspiring to grow more food without additional ecological destruction, or achieving some related technological transfers here and there, is not going to stop growing numbers of people from taking what they need, believe they need, or plain want: clearing more forests and grasslands, moving up slopes, overgrazing rangelands, decimating sea creatures, replacing mangrove forests with shrimp operations, or killing wild animals for cash or food or because they attack their livestock. "As populations burgeon," Paul and Anne Ehrlich observe, "desperate people will, as they already do so in many areas, invade nature reserves to kill what's edible, harvest what's marketable, and settle on any land that is farmable."[15]

And it is not only about desperate people. It is also about people seeking to increase their wealth. Let us revisit the trade chains of soybeans and palm oil. One study found that within a decade, present trends continuing, Chinese soybean imports for feed will outstrip the soybean production of the United States, Brazil, and Argentina combined.[16] How this demand—reflecting rising meat consumption in only one developing nation—can be met without more tropical deforestation or uncultivated lands elsewhere coming under the plow (for example, in Africa) is a big question mark. Chinese demand for soybeans will likely be perceived as an investment opportunity by those not overly concerned with the sustainable intensification mandate.[17] What's more, the expansion of oil palm plantations in the forests of tropical regions is directly beholden to the fact that palm oil has become a major ingredient in processed foods and also ubiquitous in nonfood commodities. The lucrative prospect of expanding oil palm plantations in the tropics in response to a growing consumer market will also likely override the call to avoid additional biodiversity destruction.[18]

Yet the most pernicious thing about this formulaic mandate—grow more food, don't damage more nature—has yet to be stated: it implies that the current damage our food system inflicts is an acceptable baseline of destruction. Hands down, however, industrial food production is humanity's most ecologically destructive activity. Yet mainstream discourses about feeding the world and ensuring food security rarely highlight or even mention the food system's earth-shattering demands on the biosphere.

Instead, the current ability to produce ample amounts of food—in principle enough for all people currently alive—appears to merit a different conclusion: that humanity's food-producing capacity is not constrained by natural limits as it is for other species; and that this productivity might possibly be stretched even further by means of managerial and technological innovations. According to this line of reasoning, technology-wielding humanity is different from all other species that are inexorably checked by nature whenever their numbers exceed the capacity of the environment to sustain them. Indeed, the stated or implicit belief that humans are exempt from any natural "carrying capacity" is a cornerstone assumption of the mission to continue expanding food production to support the coming billions.

A brief digression on the concept of carrying capacity is in order. This demographic idea refers to the maximal number of a species (or population thereof) that its environment can support, without that environment becoming too degraded to support the same species in the future. If a species *does* exceed its carrying capacity, with numbers mounting beyond what the natural setting can sustain, the penalties are severe: starvation, competition, and disease ensue until the population adjusts downward within a supportable range. (In unfragmented wilderness such penalties are often averted by members of the species emigrating out of the site where numbers have exceeded the capacity of the environment to support them. Over the course of history, human beings have often done the same.) While this natural law of the relation between population size and sustenance appears broadly applicable in the animal kingdom, here's the key point underscored regarding the human exemption: it is averred that history has shown that it does *not* to apply to humans.

In the early nineteenth century, Rev. Thomas Robert Malthus endeavored to relate the logic of natural limits, and the costs of transgressing them, to humanity. He predicted that because population size tends to grow faster than food production, human numbers would eventually outstrip the available food supply and people would suffer the consequences of famine, war, and disease. The two centuries following his analysis, however, did not see a human population crash, as food production kept up with swelling numbers of people. (During the last half of the twentieth century, the rate of food production actually outpaced the rate of population growth.) Thus, Malthus's thesis was largely laid to rest, and the doctrine of human exemptionalism from natural limits received a victorious boost.

As recently as 2013, Erle Ellis recited the thesis of exemptionalism in a *New York Times* editorial titled "Overpopulation Is Not the Problem." "There really is no such thing as a human carrying capacity," he stated, affirming a stock cultural belief. "The idea that humans must live within the natural environmental limits of our planet denies the realities of our entire history, and most likely the future. Humans are niche creators. We transform ecosystems to sustain ourselves. This is what we do and have always done. Our planet's human carrying capacity emerges from the capabilities of our social systems and our technologies more than from any environmental limits."[19] Ellis is right of course—but he omits mentioning that the main reason there have been few environmental limits to humanity's expansionism is that civilized humans have never respected the carrying-capacity needs of nonhumans.

To drive the point home with a pop-culture analogy, the Borg would also surely claim that there is no such thing as a "Borg carrying capacity" in the universe, but their proclamation would not negate the reality that the Borg inflate their enterprise by means of destroying and assimilating others. Moreover, the Borg do not think of themselves as conquistadors who make others suffer unnecessarily and diminish the cultural and biological diversity of the universe. Rather they think of themselves as a race set apart from all others, and playing out a destiny bequeathed upon them by their superlative nature. Rote denials of "human carrying capacity," while superficially true, turn a blind eye to the fact that humanity has succeeded in continually stretching its carrying capacity by means of appropriating the homes and sources of livelihood of nonhuman species, systematically destroying them in the process. Crediting human ingenuity for extending and defying natural limits masks that this feat has been, and continues to be, accomplished by taking over nonhuman nature, exterminating what is in the way with virtually no restraint, and vastly reducing the splendor of Earth's diversity.

It must be granted to critics of the Malthusian thesis that foreboding forecasts of the human population exceeding the amount of available food and crashing have not come to pass (at least to date). Malthus's forecast has been disproved by means of ecological mayhem: converting Earth's most fertile lands for agriculture after denuding them of their life-rich forests, grasslands, shrublands, and wetlands; taking over extensive natural areas for domestic animal grazing; appropriating half the world's freshwater, with the biggest share for agriculture; applying enormous quantities of synthetic chemical and fertilizer pollutants; decimating Earth's big herbivores, carnivores, and fish; and causing a systemic marine life crisis by plundering inconceivable numbers of wild fish. The fact that "there really is no such thing as a human carrying capacity," to cite Ellis again, means that *there really are such things* as an extinction crisis; plummeting populations of land, freshwater, and marine wildlife; and wholesale destruction or simplification of ecosystems and biomes.

Thus the ostensibly winning argument that humanity is uniquely capable of keeping its food production apace with, and even ahead of, its demographic growth is hollow if not disingenuous: it conceals what stretching food-producing capacity has portended for the planet. It reveals an inability to appreciate that human carrying capacity (how many people the Earth can support) has been extended not only because humans are so clever at manipulating natural processes and inventing stuff, but through forcefully appropriating nature's fertility for aggrandizing the human enterprise. Moreover, the exemptionalism thereby displayed—that humans are not subject to natural laws like other species—serves as an ideological accessory to the worldview of human supremacy on two fronts. First, by always crediting human ingenuity for defying carrying-capacity limits, the doctrine of exemptionalism masks that appropriating the breathtaking wealth of nature has bankrolled such defiance. Second, the credo of exemptionalism bolsters humanity's sense of superiority vis-à-vis all other creatures, thereby reinforcing the belief that humans are proportionately that much more entitled. The acts of war on nature that undergird expansionism—for food production in particular—thus become unrecognizable as acts war. Ditto for the detractors of Malthus and enthusiastic supporters of human exemptionalism: for them the violence that characterizes (and is required for) biosphere-scale food production is apparently a moot point.

The question of whether ultimately, down the road, there are (or not) natural limits to humanity's food-producing ability, which will (or will not) check demographic growth, is not an interesting question; the experiment required for the final verdict is an ugly one, either way. Instead, the alternative imperative of scaling down and pulling back invites a beautiful alternative: humanity can reject life on a planet converted into a food factory to sustain a maximal number of consumers and, by embracing limitations, enable the preservation and return of life's abundance, diversity, and complexity in vast expanses of the biosphere's landscapes and seascapes.

To many, the call for limiting the human enterprise has an unpalatable ring about it. This seemingly "gut" feeling is a conditioned effect of the unlimited entitlements that the anthropocentric worldview has bequeathed upon the human race. Spurning limitations for human life on Earth would strike us as odd were we to invest some thought into this matter. Are not limitations always called for in cultivating a beautiful and ethical life? Why would a beautiful and ethical life on the planet be an exception? We are profoundly aware of the need for limitations in relation to personal and social behavior—we know that "more" and "no boundaries" rarely serve. Too much stuff in the household results in cacophonous clutter. Too much food brings sluggishness, obesity, and disease. And, to borrow an analogy Bill McKibben has made, while drinking one beer in the right place and at the right time can yield a good experience, drinking a six-pack never does. Why, then, are limitations regarding humanity's relationship with the biosphere so widely seen as restricting human potential, when they mean precisely the opposite—enabling the full blossoming of human potential within a beautiful and thriving world?

The Depredations of Industrial Agriculture

"Nothing has driven more species to extinction or caused more instability in the world's ecosystems than the development of an agriculture sufficient to feed 6.3 billion [sic] people," wrote life scientist Peter Raven, a few years ago, about industrial food production.[20] Crop and animal agriculture, argues sustainability scientist Jonathan Foley, "has become the single biggest hammer we're smashing the planet with."[21] Worldwide, agriculture takes up to sixty times the amount of land of all cities and suburbs combined.[22] Thus, it is far less in our living quarters that our ecological footprint resides than in what we eat. Despite its already massive imposition on the biosphere, food production will need to increase substantially within this century.[23] If present trends continue, it is estimated that in the next fifty years humanity will produce as much food as people have consumed over the course of human history—a statement difficult to wrap one's mind around.[24]

The goodness and beauty of food cannot be divorced from the character of the sites and methods of its production. Mass-produced industrial food is neither good nor beautiful, but constitutes a destructive burden that we must move in the direction of liberating the Earth and ourselves from.[25] To do so, it is necessary to start by disabusing ourselves from the pitch that industrial food represents "progress" and hence stands as a benchmark of accomplishment to build upon moving forward.

Environmental and cultural historian Leo Marx offered an insightful analysis of the modern idea of progress in a classic paper titled "The Domination of Nature and the Redefinition of Progress."[26] There he dissected the notion of progress in a way that parts company with both its uncritical conception of history as "a record of continuous improvement" and its wholesale rejection as a historical narrative leading indelibly toward ecological decline if not catastrophe. Marx drew a distinction between two meanings of progress: one with a lowercase *p*; the other with an uppercase *P*. Lowercase progress refers to developments in specific domains or fields wherein one might reasonably point to the fact that progress has been made. (For example, one can cogently claim that progress has been made in dentistry or particle physics over the last, say, two hundred years.) But when one sweepingly beholds the last few centuries, or the entire course of human history, only to discover "the march of Progress," then, Marx convincingly argued, we are in the throes of an ideology. The ideology of Progress is the blanket construal of everything as an improvement over what came before, without bothering to look at the fine-grained picture or any of the specifics. The narrative of Progress is especially shrewd in eliding mention of precious and important things that have been lost in history's unfolding—the loss of life's splendorous diversity being a most precious and important case in point.

When we hear industrial food production praised as progress, and even religiously parlayed as a "miracle," we are witnessing a manifestation of the grand narrative of Progress. This view of industrial agriculture backgrounds, and often just plain hides, everything that industrial food has devastated in its wake—precious and important things like topsoil, wild creatures, lush grasslands and forests, life-abundant wetlands, ancient aquifers, pristine rivers, biodiverse estuaries, small-hold farmers, and a virtuous ethic of farm animal care.

The methods and ramifications of industrial food belie its portrayal as progress, for its treatment of land, seas, animals, and people is ecologically unsound and ethically repugnant. Industrial agriculture and industrial fishing are comparable to the Holy Inquisition: steeled by the pseudo-gravitas of social authority, the gains industrial food production reaps from land and seas are obtained through abuse of power and inflictions of suffering. And what is extracted is always more—more raw materials, more commodities, more profits, and more bad food.

What is taken away is life, including its very ground of the living soil. Industrial agriculture degrades soil by exiling nutrient-producing farm animals from the land, by decimating soil biodiversity through chemical applications, by compacting the ground with large machinery, and by erosion-promoting reduction of plant cover and of soil organic matter in monocultures.[27] Monocultures obliterate and replace entire biomes. What is taken away is a great diversity of wild plants and animals, evicted from their homes. Industrial agriculture has driven domestic animals off the farms, crowding them into CAFOs to endure short and miserable lives. What is taken away is the livelihood of farmers. Small-hold farmers have been consistently forced off the land in both the developed and developing world by the surplus productivity, and consequent torpedoing of grain prices, of large-scale industrial agriculture. Or they have been forced into conforming to the gigantic agriculture mold of utilizing chemicals, fertilizers, machines, corporate-controlled seeds, migrant laborers, and cash-crop markets. Farmers have also paid for the ascent of the industrial food system with their lives, plagued in some places by suicide epidemics and everywhere by a higher than average suicide rate. To mention one appalling example, in the last three decades tens of thousands of Indian farmers have committed suicide.[28]

A human-scale relationship with the land—what visionary farmer Wes Jackson likes to call the right "eyes-to-acres" ratio—has been reconstructed into a monotonic gigantism etched on the land and imposed on the seas, its structure and workings testifying to an authoritarian refusal to enter into conversation with the living world.[29] What pours out of those industrial landscapes are raw materials for the industrial machine, which always figures a way to turn excess into profit. There can never be too much of more it seems, because some shrewd innovation—like "corn syrup" or "pink slime"—turns surplus into business opportunity.

This critique is not heading toward a rejection of cultivating the land. Far from it: growing vegetables, fruit, legumes, and grains is good. Scale and methodologies make all the difference. There are many aspects even of preindustrial agriculture that must be changed in making crop and animal agriculture harmonious with the natural world— most especially its tendency to deplete land fertility and move on to take over new wildlands.[30] Even so, industrial agriculture—inextricably coupled with population growth with which industrial food production is in a positive feedback loop—has brought the onslaught of agriculture to a new epiphany. Industrial cultivation upped the ante of agriculture's destructiveness by moving its footprint off the cultivated fields themselves for the first time since the invention of agriculture.[31] It introduced fossil fuels and unprecedented toxic chemicals into the enterprise, poisoning rivers, groundwater, estuaries, atmosphere, and all living beings from frogs to farmers.

Industrial agriculture is built on poisoning the foundation of food: the soil that is a living membrane of staggering and mostly unknown biodiversity, ranging from bacteria, protozoa, and fungi to nematodes, earthworms, insect and mammal life, and more.

The design of industrial agriculture is masterminded as a total displacement of ecologies and of the critters, plants, and processes that form them. Author Lierre Keith accurately describes agriculture as "biotic cleansing, drawing down species, ecosystems, and soil to temporarily increase the planet's carrying capacity for humans."[32] Industrial agriculture is as far removed from a creative endeavor in cooperation with the natural world as an endeavor can get. Mining the living tissue of the soil depletes and in the long run kills it. Subsequently, soil's continued services—as a kind of dirt platform—must be assisted by the application of another class of major pollutants: synthetic fertilizers.

Industrial agriculture moved the farm animals off the farms, so as to turn the land over to vast monocultures serving "economies of scale": meaning, producing lots of one thing for manufacturing into cheap products that are mass consumed by billions while making a few corporations and individuals opulent. The same economy-of-scale logic was promptly applied to chickens, turkeys, pigs, cows, and other domestic animals, who have been taken away from fields, sunlight, and companionship to be experimented on,[33] crated, mutilated, piled, sickened, and mass slaughtered. In CAFOs, animals are treated as objects for turning into commodities as efficiently, cheaply, and swiftly as possible.[34] Industrial agriculture treats animals with callousness and cruelty at two levels: in the injurious ways their physiques have been engineered to grow big fast and/or to produce more, and in the unnatural environments and regimes they are subjected to during their short and brutal lives. That the mass of humanity seems to countenance such treatment of farm animals by the industrial system as necessary and unproblematic, that it is even praised as "the livestock revolution,"[35] is among the most grating cognitive dissonances of our time.

Thus, out of industrial agriculture, linked forms of violence have cascaded: for after driving the wild plants and animals off the landscape, industrial agriculture proceeded to drive the farm animals off as well, in order to grow one or two kinds of crops that are mostly turned into feed for the confined animals.[36] The brutal treatments of soil, landscape, native species, and farm animals, while seemingly separate forms of violence, stem from the dominant food system. The widespread invisibility of the system's violence has been guaranteed by a supremacist worldview that makes violence toward the nonhuman domain "disappear" by means of having construed it as a nonissue.

Toward the Abolition of Industrial Food

The unprecedented impact of industrial agriculture and fishing on the living world allows for the production of so much food as to seemingly demonstrate our ability to feed billions and, with some additional techno-managerial resourcefulness, perhaps feed two to four billion more. From an ecological perspective, however, the impact of the production of so much food demonstrates the capacity to take a magnificent planet—second to none in the known universe—and all but turn it into a one-species feedlot, while mustering the arrogance to call this act of pilfering and degradation an achievement.

Author Alan Weisman sums our current Green Revolution food system as one of "fossil fuel gluttony," "river fouling fertilizers," "dependence on poisons," and "monocultural menace to biodiversity."[37] So how will the amount of food produced be doubled or more without additional damage? The mainstream strategy is to extend the productivity of Green Revolution methodologies, complemented by biotech innovations and efficiency tunings, to places they have not yet reached.

"We're told," as Michael Pollan puts it, that "we must intensify the depredations (and tradeoffs) of agriculture in order to feed a growing population."[38]

Indeed, as the global population increases, spreading the Green Revolution to keep up with population growth is the main tack of the present-day policy framework. For example, an agricultural research center was created in late 2011 in India to help "kick-start another green revolution," by developing "wheat and maize varieties that thrive in warmer temperatures and on degraded land." The rationale for this initiative is that "South Asia's population is expected to swell from 1.6 billion today [2011] to 2.4 billion by 2050," by which time "almost 25 percent of South Asia's wheat yield could be wiped out by global warming."[39] Contemplating such numbers and statistics, it does not take outstanding insight to discern that making population stabilization and reduction an immediate global priority is the sanest means for averting food crises (and consequent human suffering) *and* slowing the march of global warming (and consequent human suffering). Yet policy circles and public media have their hopes hitched elsewhere: for example, to biotech promises, such as doubling "yields of Monsanto's core crops of cotton, corn, and soybeans by 2030."[40] A chief technology officer of that corporation (as of this writing, in the process of merging with Bayer) capped the above prophesy with a pledge: "We're now poised to see probably the greatest period of fundamental scientific advance in the history of agriculture."[41] Actually, we are now poised to see the Sixth Extinction undoing our nonhuman cohort with food production as the primary driver of that undoing.

The biosphere will be profoundly impoverished whether or not biotech plans, or new Green Revolution projects, succeed in meeting demand for food in this century. As for future human generations: Is it not more likely that they will prefer our legacy to be a thriving, re-wilded living world, rather than, say, the development of Round-Up Ready Switchgrass for gassing up a colossal global car fleet?

Calls to extend the Green Revolution or create a new one are predictably cushioned by all the ecologically correct pleas for wiser uses of water, more efficient application of fertilizers, prudent deployment of herbicides and insecticides, inclusion of no-till agriculture and cover crops, and the like. Such appeals to "greening" the Green Revolution, beyond being wishful thinking, do indeed voice necessary retooling in a time of potential water wars, fossil-fuel price volatility, arable land limitations, and other crises. But making a bludgeoning food model more efficient and marginally cleaner does not the model make good. At best, it yields a world, as Rachel Carson quipped sharply decades ago, that is "not quite fatal."[42]

I have discussed the ecological destructiveness of industrial food production to submit the following: that the social mission to double or triple it bodes ill for the more-than-human world and is a gamble, at best, for future people.

The mission to provision industrial food for a ten-billion-consumer world rests on three defective assumptions. One, that future food production can be doubled or tripled without added harm to the natural world and biodiversity; two, that current food production is a normal baseline—that it does not already demand a sacrifice of wild nature far in excess of what is reasonable and just for the more-than-human world; and three—since taking by force the livelihood and lives of other species does not count as an existential or ethical issue—that there is no such thing as a human carrying capacity, and therefore no reason to discontinue the experiment of stretching it.

Rather than scrambling to solve the problem of "feeding the world" by staying on the treadmill of more—the current agronomic and biotech obsession with increasing yields—we might instead scrutinize industrial agriculture and shred its veneer of representing progress. If growing humanity's food necessitates the demolition of continental-scale ecologies; the pollution of atmosphere, land, waterways, estuaries, and living bodies; the displacement and continued persecution of wild animals; the enslavement and egregious treatment of farm animals; human alienation from the land that sustains us; and a colossal public health bill from the spreading diseases of affluence, then such food cannot properly be called nourishing. When the life-affirming act of eating entails destroying and mistreating life on a massive scale, then it follows that we must change how we eat for life's sake—including our own.

The eventual abolition of industrial agriculture is the sine qua non of a civilization that will embrace all life's thriving—wild and free for the wild ones, cared for and respected for the domestic ones, nutritionally and ethically wholesome for the human ones. This implies embracing an agro-ecological food system: one that eschews chemicals and synthetic fertilizers, rejects large-scale monocultures, interfaces creatively with wild nature, and is primarily oriented to feeding human beings locally and regionally (and only secondarily oriented to export markets). Ending an industrial and (primarily) export-oriented food system has immediate implications for our global population size.

Among other things, moving toward abolishing industrial agriculture means terminating the use of synthetic fertilizers, which, as I discuss in the next chapter, means that in a global ecological civilization the human population cannot exceed roughly 3.5 billion people. Additional considerations of restoring and rewilding vast areas of land and ocean push that "optimum" number even lower.

Moving in the direction of superseding an industrial food system and restoring a life-abundant Earth thus yields a rationally, ethically, and ecologically robust criterion for human carrying capacity—within a biosphere free of pesticide poisons, biodiversity-ruinous monocultures, dead zones and polluted rivers, farm animal torture, public health nightmares, and bloated agricultural and pharmaceutical corporations. A new culture of agro-ecological food can be built primarily around local and regional foods, organically grown, abundant in diversity of domestic plants and animals, in friendship with wild nature and its creatures, and further enriched in diversity (at the table) by a virtually infinite number of globally sourced recipes.

Reduction in global trade of food and other products does not entail parochialism or insularity, but means that products from faraway places will be more costly and hence consumed less frequently. The foundation and guiding principle of agro-ecological farming is diversification of production, so there will be no want of food diversity in this system. Agro-ecologists are focused on restoring the great diversity of domestic seeds and breeds, while the use of hoop houses and greenhouses offer ingenious ways to prolong the seasons of "seasonal" foods. The human love affair with food—not to say the necessity of diversity of foods for optimal human health—must be honored, so the reduction of global food trade is not a call for asceticism, rationing, or limited and boring diets. All menus can be seasoned with the world's spices, while goods coming from far away will be valued as luxuries to be eaten more sparingly. Savoring seasonal eating is also a way of diversifying the palate. And a formidable international recipe set can be tailored to foods that are locally and regionally sourced. Most importantly, inventive farmers swapping seeds and techniques can create a great diversity of foods at local and regional scales.

We need a global food revolution. Instead of holding onto a biosphere-wrecking food system and the demographic growth it supports as givens, let us imagine what the world could look like if we actively relinquished both. Such a world will be more beautiful and more virtuous, expansively rewilded, abundant in ecologically and ethically produced food, with the return of streams, rivers, lakes, and estuaries to being life-filled waters from which we can safely drink and in which we will delight to swim, with deforestation halted and grassland ecologies reinstated, seas once again thriving with marine life, the extinction crisis arrested, and climate change made manageable via carbon-sequestering forests, grasslands, and organic farming,[43] as well as the deceleration of emissions and needed energy transitions. We might reasonably ask: What is detaining us from pursuing such a world and creating a civilization harmonious with a living planet? Why continue heeding the broken record of the human-supremacist worldview screeching the same old note—that embracing limitations will be an affront on human greatness?

EIGHT

Welcoming Limitations

If foresight intelligence became established, many more scientists and policy planners (and society) might understand the demographic contributions to the predicament, stop treating population growth as a "given" and consider the nutritional, health and social benefits of humanely ending growth well below nine billion and starting a slow decline. This would be a monumental task, considering the momentum of population growth. Monumental, but not impossible if the political will could be generated globally to give full rights, education and opportunities to women, and provide all sexually active human beings with modern contraception. PAUL AND ANNE EHRLICH

Given that human expansionism is catastrophically diminishing Earth's biological wealth, why is expansionism not on the table for questioning, let alone abolishing? Why are the trends of more—a growing human population, burgeoning global trade, expanding economies, and infrastructural sprawl—to be accommodated or "smartified," and their adverse effects just managed around the edges? I have argued that the worldview of human supremacy, or the widely shared belief system that humans are distinguished and due special planetary privileges, condones expansionism. By the same token, this worldview imbues any proposal for contracting the human enterprise with an aura of regression and constriction of human potential: calls to end humanity's expansionism are often charged with harboring the motive of returning humanity "to the caves." Human supremacy and its rhetorical deceits must be unmasked for a good portion of humanity, and for many in leadership positions, before continued expansionism can be seen for what it is: a runaway train that is transmogrifying the living Earth into a human-user resource base and placing all complex life in mortal danger.

To become willing to challenge the trends of more and embrace another possibility of human life on Earth, we must also become willing (in the broader culture and environmental arena) to extricate our reasoning from the discursive knots discussed earlier: the widespread belief that "human nature" is causing the ecological crisis; the notion that wilderness (free nature) is a passé and/or culturally fabricated idea; and the circulating ideology that conflates ongoing expansionism with increasing degrees of freedom for more and more people. A new imagination is needed about who we are and who we can become. We need also to recall the original ontology of Earth as a planet that produces stupendous abundance of life-forms, populations of living beings, biological processes, living phenomena, and diverse forms of consciousness, all scaling up to a shared experience of well-being, beauty, and transcendence.

Halting expansionism entails reversing, not accommodating, the trends of more: working to stabilize and reduce the global population, prioritizing robust local and regional economies over global trade, downsizing the global economy, and substantially limiting and undoing the sprawl of industrial infrastructure. The vision inspiring these proposals is that humanity should not be the sea within which remnant patches of wild nature are as islands, but the other way around: wild nature can become the vast terrain within which human societies are nestled in reciprocity with nature's abundance.

The pathway toward an ecological civilization will be challenging to forge, but it is not difficult to envision. It entails scaling down humanity's impact while at the same time pulling back our excessive presence from, and interference with, the natural world. Scaling down calls for drastically reducing excessive consumption, which requires, along with other actions and measures, lowering the global population, de-industrializing food production, relocalizing economies, and lessening global trade. Pulling back is the life-affirming project of restoring, reconnecting, and rewilding large expanses of land and ocean, so as to share the planet generously with its millions of life-forms and renew its life-nourishing vibrancy for all earthlings. Scaling down and pulling back comprise the conjoined strategies for creating an ecological civilization within a biodiverse Earth. Pursuing these strategies is explored in this chapter and the one that follows.

Facing Mass Consumption

Mass consumption is humanity's present-day hegemonic way of life: people or nations not part of this way of life are slated, and appear mostly willing, to be brought into its fold. Not to put too fine a point on it, mass consumption involves devouring vast amounts of living and nonliving matter while spewing out vast amounts of waste. Mass consumption includes the societal use of energy and materials in extraction and production processes, such as industrial agriculture and fishing, industrial forestry, mining, and manufacturing; the societal use of energy and materials for public works such as electrification, construction, transportation, and infrastructure development; and people's consumption of commodities from chopsticks and clothing to electronics and cars, as well as the consumption of foods and beverages. Mass consumption is a way of life, because it assumes the authority and prerogative to source colossal amounts of raw materials and manufactured goods (for societies or individuals) from wherever they can be sourced and by any means necessary.

This way of life proliferates never-ending services (such as globally burgeoning fast-food outlets and big-box stores) through which mass consumption can be performed and magnified. The combination of growing numbers of people, coupled with the rising global middle class participating in the global economy, means that mass consumption is not growing in linear fashion but accelerating.[1] In lockstep, severe impact on the biosphere is also accelerating, notwithstanding some efficiency gains in production, transportation, and other systems (such as savings in material throughput and energy use). Efficiency gains tend to be overwhelmed by increased use of commodities and growing size of commodities (like bigger cars and refrigerators).[2] Such gains can thus be compared to stepping softly on the brake with one foot, while leaving the other slamming down on the accelerator.

Reducing the huge demands and resulting waste output of global mass consumption will require stepping up an array of actions and policies already pursued to one degree or another: efficiency gains and conservation in energy consumption; shifting from fossil fuels to a new energy system utilizing distributed solar, wind, and other renewables; phasing out extractive industries while advancing recycling industries; reducing and, where possible, eliminating waste; transforming a throwaway economy to a circular (cradle-to-cradle) economy; abolishing destructive subsidies such as those to industrial agriculture, industrial fishing, and the fossil-fuel industry; and drastically shrinking the production and consumption of ecologically destructive foods, especially animal products (including fish). Work in all these areas is important, and while some is already under way, it can be sped up via grassroots campaigns, institutional reforms, business initiatives, and governmental policies.

Along with these approaches, the indispensable approach to lowering consumption on all fronts is stabilizing and gradually reducing the global population. As population expert Martha Campbell writes, "population is the multiplier of everything we do and everything we consume."[3] "Overpopulation," states author Patrick Curry, "has the peculiarly vicious result that simply by force of numbers, the most natural human activities relating most directly to survival and the continuation of the species—finding fuel, shelter, growing food, procreation, excretion and so on—themselves become pathological: direct threats to personal survival and to that of the species."[4]

Stabilizing the human population and embarking toward its substantial reduction are necessary for preserving the richness of the biosphere, for at least three reasons. First, whatever gains in consumption reduction are achieved by efficiency measures, recycling operations, dematerialization (i.e., less throughput in production activities), or energy conservation (in vehicles, construction design, or via behavioral changes) tend to be undermined by sheer growth in human numbers. (For example, if the average American consumes 20 percent less meat in 2050 than at the turn of the twenty-first century, total American meat consumption will still be five million tons greater in 2050 simply because of population growth.[5]) Diverse actions and policies to counteract overconsumption should of course continue to be pursued, since the global situation would be all the worse without them. Yet the positive effects of the above listed measures will become palpable once the population factor underlying consumption is stabilized and then diminished.

Second, huge numbers of people across the globe, both in the developing and developed world, overwhelm the natural world with their demands. In poor nations, high population density and rapid population growth drive forest, grassland, and wetland destruction, often for subsistence food production and increasingly for large-scale, export-oriented agriculture. As researchers Jeffrey McKee and Erica Chambers have found, "human population density is a primary cause of biodiversity losses, in a large part mediated by agricultural land use, and is thus a key factor that must be addressed to reduce future threats to Earth's biodiversity."[6]

Examples of destructive impact due to high density and rapid growth in the developing world include Haiti, Ethiopia, Niger, Ghana, Madagascar, Pakistan, and the Philippines—all nations that have destroyed vast portions of their forests in recent times. For instance, even until the middle of the twentieth century, Pakistan was more than 30 percent forested, but its forests have dwindled to 4 percent of its territory in our time.[7] (Pakistan has about two hundred million inhabitants, and, despite a looming water crisis, is projected to grow to 395 million by 2050.[8]) On a similar note, the profiles of rapidly growing African nations, as summarized in the open-access World Fact Book, reveal a near identical litany of environmental disasters, most especially in countries with soaring fertility rates: deforestation, desertification (from overgrazing), poaching, soil erosion, and pollution. Pressures of high population density on biodiversity are not foreign to the developed world, as the eastern coastline of the United States and the state of Florida, among other places around the world, attest.[9]

Meanwhile, hundreds of millions of consumers in the developed (and increasingly urban developing) world also destroy (and have destroyed) much of the biodiversity of their land and sea territories. Adding insult to injury, they have the power to sponsor more destruction remotely. For example, deforestation of Brazil's Amazonia, Indonesian forests, and the Congo basin is beholden to pressures from rich nations, via a global trade regime that those nations have favored and established. Economic factors associated with mass-consumption markets in countries with which developing nations do business (such as the United States, China, Japan, and European countries) drive habitat destruction and species extinctions.[10] Importantly, the developed world is also historically most responsible for greenhouse gas pollution, thereby instigating a climate-change episode that if allowed to continue unabated has the potential to bring about a holocaust for all complex life on Earth.[11]

In brief, the second reason that numbers of people in developing and developed nations need to be reduced is that they collude, in different as well as interacting ways, to diminish biodiversity and fuel the extinction crisis: the poor via high densities and rapidly growing numbers in the vicinity of wild places and consumers by the energy, food, mineral, and other demands at home and abroad of their ranks swelling into billions.

The third reason that a far lower global population is called for is connected (as earlier argued) with the imperative to move in the direction of abolishing industrial food production. The organic, local, fair, and slow food movement has made significant strides toward revamping the food system to be Earth friendly, as well as animal, farmer, and community friendly, and to offer a high-quality nurturing diet that should become universally available and not just the privilege of the wealthy. The question of whether organic food production could be scaled up to feed the present and coming population has often been posed in recent years.[12] While it may offer an interesting agronomical inquiry, from the perspective of humanity inhabiting a biodiverse planet it is the wrong question. The pertinent question is: What must the global population size be in order for all people to be well fed on organic, diversified, and mostly locally and regionally grown food, while also allowing terrestrial and marine space to be freed and rewilded? Simply replacing industrial monocultures with organic production systems, while promoting less polluting and healthier options, would not enable the reduction of land-use under cultivation.

The solution lies in pursuing the goal of gradually lowering the global population to a level that can be sustained by an agro-ecological food system that would also not place huge demands on land and seas. This calls for shifting perspectives on human carrying capacity away from the standard definition of the *maximum* number of people the planet can adequately feed, clothe, and shelter, toward formulating a robust touchstone: the number of people that an ecologically sound food system can support while simultaneously allowing for the flourishing of wild nature. Since the dominant food system—founded on industrial animal and crop monoculture production, primarily designed to serve global markets—is the most catastrophic force on the planet, it follows that addressing life's crisis requires revamping the food system.

Revamping the food system, in turn, has implications for global population size. Humanity's current and growing numbers are beholden to industrial food production. It is estimated that roughly 40 percent of our population could not be sustained without synthetic fertilizers; otherwise put, synthetic fertilizers have nearly doubled human carrying capacity.[13] Between 1961 and 2000, fertilizer use increased 700 percent.[14] The global population is 7.5 billion people and growing primarily because industrial food has made this possible. Industrial food production on land and seas has thus underwritten exponential growth, enabling more people to be alive, all at once, than would otherwise be possible.

Moving toward a food model that is organic, diversified, and locally and regionally oriented, using no synthetic fertilizers and chemicals, among other changes, means that human numbers must be gradually lowered to at least half present levels. Considerations of land and sea protection push the number even lower than half its current size, for in order to restore a life-thriving planet large areas of habitat must be returned to wild beings and ecologies. In addition to the mandates of an organic polycultural food model and large-scale protection and restoration of wild nature, it is also imperative that people inhabit places nearby where food can grow for both food security and ecological reasons. All considerations in hand, a global population closer to two billion people is an initial ecologically sound and rational goal, enabling the conservation of a biodiverse planet, a connected global civilization, a high-quality and equitable standard of living for all people, and the co-flourishing of humanity and the living world. It is also a feasible goal as I argue below.

The argument in a nutshell is this: If we choose to change how we eat—so that food production ceases impoverishing the biosphere, while displacing, killing, and mistreating myriad beings—then we must lower our population. The proposal that the human population should not exceed the support capacity of an organic agriculture was also made by ecological economist Nicholas Georgescu-Roegen. To my knowledge, he did not develop this proposal into a full-blown argument, but in 1975 he wrote: "Mankind should gradually lower its population to a level that could be adequately fed only by organic agriculture. Naturally, the nations now experiencing a very high demographic growth will have to strive hard for the most rapid possible results in that direction."[15] The rationale underlying this proposal encapsulates an ecological bottom line for the population question, because the creation of a food system in friendship with the natural world sets relatively robust parameters for population size. Honoring such limits to population size is not a constriction of human possibility and freedom, but a necessary dimension for achieving a higher quality of life for all people and freedom for all earthlings.

This argument collides head-on with the specter haunting the population question: for many, even mention of the word "overpopulation" is taken to imply blaming the global South (where population growth is largely occurring) for the world's ecological woes. Indeed, the population issue has been so bogged down by political controversy, acrimonious debates, and knee-jerk attributions of the shady motive of "population control" that it has become, as Julia Whitty puts it, "the last taboo."[16] We can move beyond the historical baggage and mistrust surrounding the population issue by rethinking it. In this spirit, I offer four ways of reframing the global population question as a paramount issue that a unified humanity must face.

Reframing the Population Question

The first reframing involves moving beyond the prevailing quandary of whether it is excessive consumption or an unsustainable population that underlies humanity's impact, and recognizing that "overconsumption" and "overpopulation" are not distinct variables. While the celebrated IPAT formula—of impact as a factor of population, affluence, and technology—has been useful analytic shorthand for elucidating the big factors underlying ecological damage, it has also encouraged relegating "population" and "affluence" (i.e., high consumption levels) into separate explanatory silos. From this balkanization of nature-impacting variables, it was an easy move to regard the global North (the developed world) as having a "consumption problem" and the global South (the developing world) as having a "population problem." This assessment became, and to some extent remains, a truism.

The truism, however, is muddled and inaccurate. For starters, there are not two big factors driving nature's destruction but one—namely, overconsumption. Humanity uses the world excessively both as a source of materials from the living and inorganic world and as a sink for wastes, such as garbage, nitrogen, herbicides, greenhouse gases, confined livestock manure, plastics, sewage, and so forth that the natural world cannot absorb. Overconsumption refers to a scale of impingement that damages the biosphere, often irreparably, by using it excessively as source and sink. An enormous yet oddly overlooked variable underlying excessive consumption is numbers of people. When population size is recognized as a major contributor to excessive consumption, it becomes clear that one essential strategy for scaling down the effects of overconsumption—be it of Americans and Chinese carbon-loading the atmosphere, or of one billion people in the developing world relying on wild meat[17]—is to lower human numbers everywhere.

Of course, consumption patterns differ markedly around the world. A New York City stockbroker, a Greek cab driver, a Mongolian herder, a Russian oil magnate, a Bangladeshi farmer, a Chinese party member, an Inuit hunter, a Thai fisherman, a Brazilian cattle rancher, a South African miner, a Japanese accountant, a North Korean housewife, a Syrian refugee, and so on differ profoundly both in what and how much they consume. Vast global disparities and inequities remain. Yet at the same time, such world citizens are increasingly consuming the same commodities—for example, fast-food meals from identical fast-food chains or identical brands of apparel and cellular phones. They are also increasingly living with electricity; regulating indoor temperatures; buying cars, electronic devices, and luxury commodities; and choosing global destinations for their vacations. All these trends, of "a commodity-intensive, high-consumption existence on the model of the United States,"[18] speak to the rapid ascent of the world's middle class.

In fact, the global middle class has grown by hundreds of millions of people in just the last two decades. This is in good part due to the growth of emerging economies such as China, India, Brazil, South Africa, and others, but the trend is universal and will see more of Asia, Africa, and Latin America joining the bandwagon. This trend has outmoded a crisp dichotomy between global North and global South. While an enormous divide remains in the consumption levels of the richest and poorest quintiles of the global population, it is also the case that the middle "quintiles" have been burgeoning—a trend that will continue briskly, barring the unexpected.[19] A global middle class of roughly 3.2 billion people (2016 levels) is projected to balloon to roughly 5 billion by 2030 (with much of this growth still coming from Asia).[20] For example, by 2050, 40 percent of India's population is expected to join the ranks of the middle class, adding roughly 580 million consumers to the global economy, up from 50 million in 2006.[21] While the emergent growth of a consumer class is occurring unevenly in different parts of the world, the trajectory is global. Africa is forecast to increase between (a low of) 3 and (a high of) 6 billion people by 2100, from 1.2 billion people today.[22] As Africa (along with more of Latin America and Asia) follows the current Asian trend in the growth of its own middle class, the blow to the biosphere will be nothing less than staggering.

It has become anachronistic to cling to a North-South distinction that globalization has permanently destabilized and turned into a moving target. Indeed, the most pervasive policy-defining trend of our time is the mainstream mandate to bring a consumer standard of living to the global populace. A "consumer" is one whose material lifestyle is defined by the Western modern norm and all that norm entails—the good, the bad, and the ugly. Environmental analyses define a consumer as anyone, anywhere in the world, who participates in the money economy; has some level of expendable income; lives in electrified quarters; can afford to regularly eat meat, fish, and other animal products; drives an automobile; owns sundry appliances, electronics, and run-of-the-mill clutter; and is generally immersed in a commodified existence.[23]

The socioeconomic mandate of international policies to disseminate a consumer standard of living has resulted in the global North penetrating the global South, with the ranks of the global middle class swelling. The consequent per person increase of consumption levels has turned into a tsunami by sheer numbers of consumers in the world, immensely and dangerously accelerating ecological deterioration on land and seas. These trends of expanding consumption and escalating ecological deterioration are showing few signs of slowing. Since some level of a consumer standard of living—at this historical moment—reflects a desirable lifestyle for the global mainstream, disseminating a moderated version of such a lifestyle for all while also preserving a biodiverse planet logically demands a lower global population. As population expert Robert Engelman expresses this idea, "since all descendants of low-income, low-consumption populations . . . expect consumption-boosting economic development, a lower future population would mean less pressure on climate, environment, and natural resources by future generations."[24]

In brief, the first reframing of the population question invites us to relinquish a conception of "overconsumption" and "overpopulation" as distinct factors of impact. On the contrary, population size is a leading cause, and ever the magnifier, of the actual physical driver of humanity's impact—excessive consumption. Stabilizing and gradually lowering the global population will enable reduced consumption of everything and of corresponding waste.

The second, equally important reframing of the population question is a corollary of the first. The widespread belief that overpopulation is a "developing world problem" must be jettisoned. Overpopulation on one Earth inhabited by a globalized humanity is a global problem, full stop. It is undoubtedly the case that there is a pressing population stabilization imperative in countries where fertility rates continue to soar, exponential population growth is still occurring, and there is an unmet need for family-planning services and modern contraceptives. (Over two hundred million sexually active women around the world who want to avoid pregnancy do not yet have the effective means.[25]) That, however, should not lead us into thinking that it is poor countries that have a population problem that rich ones have solved. This erroneous but commonplace impression stems in part from the aforementioned truism: that the developed world has an overconsumption problem that—mysteriously—is supposed to have nothing to do with its population size.

With numbers of consumers in the billions and rapidly growing, population size aggregates the effects of all people's middle-class standard of living, from Johannesburg to London, from Beijing and Dubai to San Francisco and New York City, and from Vancouver to Buenos Aires. It is most especially the aggregated effects of the world's consumers—connected via global trade chains—that materially translates into colossal ecological damage. While behavioral changes, such as eating meat and fish sparingly (or not at all) or driving electric cars, can be encouraged to reduce overconsumption, the developed world and emerging economies will go a long way toward scaling down their impact by steadily lowering, or continuing to lower, their total numbers to levels well below the present ones.

In other words, there exists no compelling reason to black-box the populations of developed countries and of emerging economies as "normal" simply because they have reached near replacement, or below replacement, fertility rates. Countries like Germany, the United Kingdom, Australia, and the United States, for example, are overpopulated by at least two criteria: one, the higher quality of life their citizens could enjoy of co-thriving ecological and human communities within their borders, if their populations were lower; and two, the higher moral standard they would achieve by eliminating or (for starters) substantially reducing their hefty ecological footprints abroad, if their populations were lower. Reduced populations in developed countries translate into reduced need to produce and import food and nonfood commodities, as well as expanded protected areas of wild nature, more urban green space, less traffic, less sprawl, fewer infrastructure requirements, fewer strip malls, and reduced pollution. As anthropologist J. Kenneth Smail nails the broader point, "only a global population 'optimized' at a considerably reduced size will provide the opportunity to build a much better quality of life for everyone."[26]

In sum, population size is not strictly a developing world problem, but a global issue and charge. One of the most effective and tangible ways to address climate disruption, as well as to curb the excessive consumption of everything, from food to cell phones, is to move toward the substantial reduction of the number of people worldwide, including the populations of the developed world and emerging economies.

The third reframing of the population question underscores that overpopulation is both an ecological and a social justice concern. Every human being on the planet should have ready access to state-of-the-art reproductive health services, free or affordable contraception, responsible and accurate information about fertility, counseling support, and comprehensive sexuality education. The institution of child brides, no matter how "culturally customary" it may be, should not be tolerated by the international community any more than slavery is tolerated as culturally customary. The practice of marrying off girls must be decried for what it is: the sexual abuse of children, adolescents, and young women who are subjugated by social practices and cultural mores into lives of full-time breeders. Making family-planning services fully available and promoting global gender equity are urgent international priorities that will hasten the eradication of this injustice.

Additionally, the population question, reframed in the geopolitical moment of the twenty-first century and beyond, must have no truck with coercing people into having fewer or no children. Wherever, in the name of population control, such coercion has occurred in the past, the historical lessons can be heeded and similar policies safeguarded against. Regarding the intersection of coercion, family planning, and reproductive rights, demographer Malcolm Potts points out that "past episodes of coercion have cast a long shadow over international family planning and must never be repeated." Crucially, he goes on to add that "the new imperative is to ensure also that women are not subject to coercive pregnancies because they are denied access to the information and technologies they need for the voluntary control of childbearing."[27] Thus, not only must governments and family-planning policies disavow coercing people into childbearing decisions, but the proliferation of family-planning services must also be applauded as playing a vital role in rescuing girls and women from coercive, high-fertility norms in whatever sociocultural enclaves these occur.

Feminist critiques of population control have denounced past coercive reproductive policies as demeaning human dignity and violating human rights.[28] "In the 1970s and 1980s," writes public policy expert Ellen Chesler, "high-profile abuses of human rights by numbers-driven population programs, especially in China and India, undermined a well-established consensus that family planning programs are an essential tool of sound public health and development practice."[29] Such coercive policies have been censured as illustrating that overpopulation concerns and demographic thinking can become "tools of tyranny," in the words of feminist author Michelle Goldberg.[30] Yet episodes of authoritarian population policies—such as forced sterilizations and top-down enforcement of family size—do not demonstrate that concerns about overpopulation should be laid to rest: rather, such human-rights violations are historical lessons of how not to design population policy. Instead of drawing this rational conclusion, the backlash to certain coercive national policies contributed to the population question and voluntary family planning slipping off the international agenda.[31] As Campbell notes, a "false generalization" equated family-planning programs with coercion.[32] This has been a profound misfortune across the board—for the natural world, for children's and women's rights, and for future generations whom we are burdening with unimaginable challenges.

Unfortunately, certain feminist and other critics of "high-profile" population policy abuses fail to be as vociferous about the "low-profile" abuses of countless anonymous girls and women subjected to socio-cultural and/or religious patriarchal norms of bearing many children, with little say in the matter, and often starting their childbearing lives barely out of childhood themselves. While undoubtedly there are, and always will be, women who genuinely want many children, this is an outlier predilection. Decades of research have revealed that women's natural proclivity lies in the opposite direction. "When women are offered modern contraception in respectful ways and supported with correct honest information, and backed up by safe abortion, then the birth rate always falls."[33] Fertility declines follow from a straightforward bio-cultural cause: that the overwhelming majority of women, when they attain the means to control their fertility and achieve free choice, rarely want more than one or two children (if any), because numerous offspring are hard on the female body and also take away time from personal pursuits.

Women's inherent propensity for few children surfaces straightaway once barriers to reproductive self-determination are removed.[34] Because of the universality of this trend, Campbell speaks of women's "latent desire" for few children—an observation that, given the heightened mortality risks of pregnancy and birthing, makes complete biological and evolutionary sense. Unsurprisingly, as Engelman notes, "the key to trimming family size is a consistent focus on improving women's lives, including economic opportunities and legal guarantees" equal with those of men. "Despite perceptions to the contrary," he adds, "national economic growth alone does not push fertility down powerfully."[35]

It is anything but coincidental that high fertility rates today persist in societies where women have little economic power and property rights, limited access to education, and often no say in family planning or even sexual contact. As a 2007 UK parliamentary study put it, "no woman can be free unless she has the technologies and information required to enable her to decide whether, and when, to have a child and to escape the tyranny of unintended pregnancy. Women with numerous pregnancies and life-long child care find it difficult to participate in education, markets, or politics."[36] Strong activism for family-planning services is, in itself, a powerful tool for subverting social structures that subjugate girls and women.

As research over the past decades has shown, reduction in fertility rates is closely connected with the education of girls and women, and ultimately with the creation of sociocultural environments within which all women can pursue self-realization and exercise the freedom of whether, when, and how many children to have. The population problem today must be reframed in terms of the attainment of human rights: to free girls and women from social structures that define for them, and impose upon them, the chief identity of motherhood. Nor does this aspiration of freedom apply only to the developing world. It is also about rescuing disempowered inner-city and rural girls of developed countries from lack of opportunity and low self-esteem, by means of nourishing their talents, empowering them through education and support, and making fully transparent their choices about childbearing and sexual contact. When women are empowered, they make their own reproductive decisions that almost invariably mean having few or no children.

I turn to a fourth way and final suggestion for reframing the global population question. It comes cued in an anthropocentric concept that nonetheless serves as useful shorthand here: Livestock. The global population of livestock today is over tenfold the global human population. By 2050, present trends continuing, the world's livestock will consume as much food as four billion people.[37] Given the comprehensive ecological calamity that cattle, pigs, sheep, chickens, and other farm animals cause—from vast amounts of land use and freshwater diversion to fouled streams and rivers, and from deforestation and dead estuaries to global warming—some sustainability advocates have averred that "we do not have a human population problem, we have a *livestock population problem*."[38] Even as this argument points in the direction of a quasi-valid point, it is misleading.

The biosphere does indeed have a livestock population problem, especially because the population size of human beings—who are omnivorous by nature and (today) certainly by inclination—has vastly outgrown being harmoniously proportioned with the living Earth. It is the case, of course, that if all people ate a plant-based diet, humanity would cause considerably less harm to the planet. Indeed, "eating as few animal products as possible—ideally none—is a powerful way to be part of the solution."[39] Recently, George Monbiot made a strong, succinct argument for a vegan diet. "It's not hard to see how gently we could tread if we stopped keeping animals," he writes. "Rainforests, savannahs, wetlands, magnificent wildlife can live alongside us, but not alongside our current diet."[40] It is also the case that if meat, dairy, and other animal products were made more expensive—with their externalities folded into their pricing—their consumption would be discouraged and the livestock population might then become globally reduced.

In the world we live in, the first possibility of a global majority of vegans is (for the foreseeable future) improbable. (Efforts, however, on the part of prominent individuals and institutions to promote a plant-based diet are desirable, both for the immediate health benefits they bring to those who adopt it and for their contribution to changing the cultural politics of food in the longer term.) The second possibility of increased prices of animal-derived foods would impact lower-income groups far more than those in middle- and higher-income brackets. On the latter point, today's huge educational and income inequalities already fly in the face of food justice: educated people eat fewer (or no) animal products; the wealthy are able to purchase higher quality (organic and local) animal products that are a lot more expensive than the mass-produced ones; and poor and uneducated people consume fast-food junk at McDonald's-type establishments—while the fast-food industry specifically targets and markets to lower-income communities and to their children and is swiftly globalizing its seductive, disease-causing operations.[41]

How can we move from the current livestock-heavy food system that is an ecological disaster, socially unjust, engendering an enormous and global public health crisis, and cruel to animals who live and die in factory farms? To move in the direction of a global diet that is, simultaneously, realistic, diverse and healthy, ecologically friendly, socially just, and kind to animals, the empirically sound approaches include substantially lowering the human population so that the livestock population can follow suit; universally educating people to eat fewer animal products, effectively enabling the further reduction of livestock numbers and the betterment of public health; and returning all farm animals to farms where they belong for their own sake and for the sake of enhancing the fertility of cultivated soils. Of course, getting from where we are to such a world is a tall order, but at least it is a realistic tall order, which also includes everyone's well-being. People, animals, and wild nature will benefit and can ultimately thrive in a world of far fewer people, equally educated to sparingly eat animal-derived foods that are nutritionally, ecologically, and ethically virtuous. Recent research by ecologist Brian Machovina and his colleagues indicates that the restoration of a biodiverse biosphere demands reducing animal product consumption to roughly 10 percent of the human diet.[42]

To review, I have suggested four ways of reframing the population question: one, as a major variable undergirding excessive consumption and waste; two, as a global issue that a unified international community must address by promoting policies that will stabilize and slowly reduce humanity's population; three, as a matter of human rights (most especially children's and women's), which begs for the achievement of full gender equality everywhere in the world; and fourth, as tied up with the present-day livestock calamity, which is devastating biosphere, animal lives, and human health.

The purpose of rethinking human population along these lines is to contribute to seeing overpopulation in a new light that eschews past barriers to a much-needed conversation and moves us toward discerning the rationality of stabilizing and lowering our numbers. I have demonstrated that population size significantly drives excessive consumption and waste and is not a variable independent of overconsumption. In light of a globalizing world and the mainstream international policy of lifting people out of poverty into middle-class lifestyles, I have also urged the recognition that overpopulation is a global problem and not a developing world one. It is necessary to counter the silence that has surrounded population size and growth in scientific, policy, and public arenas in the last two decades, while simultaneously endeavoring to promote broad agreement about the desirability, for human beings and the biosphere, of fewer human inhabitants on Earth.[43]

Reducing the Global Population by Enhancing Human Rights

An optimal global population figure remains elusive, because the impact of human numbers is connected with societal lifestyles (overall consumption patterns and waste output) and with technological systems (especially in the energy, food, transportation, and electronics industries). Thus what an optimal global population will be, say, in the twenty-second century and beyond is difficult to foresee, for it depends on the material standard of living people will gravitate toward,

the food system and diet they embrace, and technological developments and shifts virtually impossible to predict. From the present vantage point, however, a compelling initial approximation is two billion.
This "optimal" figure was first estimated by scientists Gretchen Daily
and Paul Ehrlich who argued that a population of roughly two billion
would enable *both* the conservation of vast tracts of wild nature and a
culturally diverse and interconnected global civilization.[44] Agriculture
expert David Pimentel arrived at the same ballpark figure for a global
population that organic agriculture could support.[45] Thus, from our
present perspective, two billion is a sound ideal that, along with other
indispensable shifts, could support the co-flourishing of natural world
and humanity.

Making a U-turn from the current ten to eleven billion trajectory
toward gradually reducing our numbers over the next few generations
is more feasible than often believed. Lessons from successful population policies in the twentieth century—in places with cultures and re-
ligions as diverse as Tunisia, Thailand, Bangladesh, Costa Rica, South
Korea, Kerala (India), Cambodia, and Iran—reveal that the most effective trans-societal strategy for lowering fertility rates is the launching
of a comprehensive, well-designed, well-funded, and human-rights-
grounded campaign toward that purpose. "Defying the expectations
of economists," notes Engelman, "countries didn't even need to get
wealthier to become less fertile."[46] Women respond positively, and
wherever such concerted campaigns have been implemented fertility
rates decline relatively swiftly.

The world is demographically diverse. Many regions are experiencing significant and rapid growth, while developed countries and a
number of emerging economies have entered a phase of low to negative
growth. Despite the extreme diversity of demographic profiles, a common thread links the policies that will help promote an ecologically
thriving planet and high-quality human life. The common thread includes three development requirements that we might rightly regard
as configuring a universal human birthright: accessible and affordable family-planning services, including modern contraceptive technologies; educational opportunities for girls and women and, more
generally, women's empowerment; and what the United Nations calls
"comprehensive sexuality education," beginning early in educational
curricula and continuing throughout children's and young people's
schooling, with age-appropriate and age-pertinent material.

In unison these three developments—that all improve human
lives and empower people to make well-informed and well-supported

life choices—will ensure a world in which small families will be the norm. It is vital for the integrity of the biosphere and a hope-filled human future that state-of-the-art family planning, full gender equality, and comprehensive sex education become part of every society on the planet.

High priority on the international agenda must be for all people to have access to, and unhindered agency to use, affordable or free voluntary family-planning services. The aim of family planning is defined as supporting the right to choose whether and when to have a child by providing the means of implementation, while promoting "voluntarism, informed choice, rights, and equity."[47] Achieving universal family planning calls for sustained commitment, for it demands the investment of substantial financial and human resources. Ensuring the viability of family-planning programs means building and equipping clinics (or maintaining/upgrading existing ones), schooling and staffing programs with experts (doctors, social workers, psychologists, and so forth), training health workers as grassroots consultants, making the broad spectrum of modern contraceptives easily available in stores as well as through other delivery systems, and subsidizing a good portion of much of the above for people who require financial assistance.

"The crucial ingredient for success," according to population scholar John Bongaarts, "is political will and a commitment to family planning at the highest levels of national and international policymaking."[48] As population expert Madeleine Fabic and her coauthors state, "renewed commitment, policy development, financial support, and increased attention are required to ensure that family planning is an integral part of the post-2015 agenda." With global effort, the authors add that the greatest portion of demand for family planning can be met, in developed and developing countries, in as few as fifteen years.[49]

International funding has declined markedly in recent decades even though the financial backing to bring family-planning services that allow women to make their own decisions has been pivotal in countries where fertility rates have fallen. Reversing this financial shortfall is crucial. As of 2016, only 1 percent of overseas development assistance is allocated to family planning. Bongaarts points out that even a mere doubling to 2 percent, along with strong support of developing country governments, would make substantial progress possible. Concerning the developed world's responsibility, it is reasonable to insist that rich nations and institutions provision the financial backing and expertise for bringing reproductive health services around the world—including their own home territories. (For example, half the pregnancies that occur in the United States are unintended, a statistic that speaks to a social, cultural, and educational failure, not simply to a weakness of human nature. I discuss the connection between unintended pregnancy rates and population growth below.)

As stated earlier, over two hundred million women have unmet need—even though they desire the means to control their own fertility, they have no access to them. Moreover, there are huge discrepancies of access to family-planning services between *and* within the developing and developed worlds. Overall, the wealthiest, most educated women tend to have the resources and freedom to avail themselves of family-planning support, while economically and educationally disempowered women are deprived of this critical decision-making service.[50] Regarding family planning (along with other human rights), the motto of the status quo appears to be "to those who have it shall be given and to those who do not have it shall be taken away." It is high time that every woman in the world can avail herself of family-planning facilities and expert support, and be free of oppressive sociocultural norms that prevent her from consulting such services.

Millions of poor women face a gauntlet of barriers to reproductive self-determination. Barriers can be physical (no family-planning support available); cultural (dominant patriarchal norms for multiple offspring, usually along with early female marriage and lack of educational opportunities); educational (no sound information and/or wrong ideas about contraceptive technologies, and no viable alternatives to marriage and childbearing); religious (the belief that God determines how many children one bears and not one's choices and behaviors); and often a combination of the above.[51] When barriers are eliminated and women become personally empowered, fertility rates nosedive. Indeed, family size may fall "even in poor and illiterate communities once the many barriers that bar women from access to the technologies and information they need to separate sex from pregnancy are removed."[52]

Responding to population growth with the seriousness it deserves will not only soften the blows of ecological challenges but decidedly advance human rights, especially women's and children's rights. Wherever women are empowered educationally, culturally, economically, politically, and legally, fertility rates fall.[53] Populations tend to move toward states of zero or negative growth when women achieve equal standing with men, as long as family planning (including contraceptives and sound information about their use) is readily available. Education is a key variable. While other factors play important roles, the number of years of a girl's and woman's education varies, on average, exactly inversely with the number of children she will have.

Making education for girls and women an ambitiously pursued international policy is not only laudable in itself, but pivotal for the future of the global population. The level of a woman's schooling is a reliable "predictor" of her fertility. Worldwide, women with no education have an average of 4.5 children each, while women with a university education have an average of 1.7. A declining fertility rate tracks increasing years of schooling in between.[54] The connection between low fertility and female education is so robust that a 2011 Vienna Institute of Demography study that modeled different educational scenarios found that if every country invested enthusiastically in schooling girls, by 2050 there could be a billion fewer people than if nothing changes.[55] The significance of female education is evident in another recent statistic from Africa. African women with no education have, on average, 5.4 children; women who have completed primary school have 4.3 children, while a significant drop to 2.7 occurs with completion of secondary school; for those who go to college, fertility is 2.2 children per woman.[56] Such statistics signal that ensuring educational opportunities for girls and women will steadily move the world toward a smaller population.

Achieving full gender equality would, in all likelihood, eventually lead to global fertility below, and possibly well below, the replacement value of roughly 2.1 children. Indeed, the population question is urgent in countries where polygamous, fundamentalist, and military cultures are keeping girls out of school and women disempowered, thus adding roadblocks to a restored future. Nor does a slowly unfurling process of modern development automatically guarantee lower fertility (as is often assumed). "There is no empirical evidence that all countries and regions will drift in some magic way to a two-child family and then live happily ever after," write population experts Martha Campbell and Malcolm Potts. "Indeed, anyone who has glimpsed the patriarchal cultures found in Afghanistan or Northern Nigeria would suggest the empirical evidence is the exact opposite. Such regions are likely to go on having large families unless a massive effort is put into helping women achieve the autonomy they deserve."[57] In patriarchal societies, it is typical for men—themselves sadly indoctrinated into and deluded by an unjust belief system—to believe that childbearing decisions are theirs alone to make. These are also the societies where partner sexual, physical, or emotional violence (sometimes in response to a woman expressing interest in contraception) is more prevalent.[58] In a global-izing world converging toward the universal moral precept of equality between all people, patriarchy needs to be fast-tracked out of existence.

Alongside establishing family-planning programs and agitating for women's equality, the third human-rights development for a world of small families is instituting comprehensive sexuality education in curricula around the world. Comprehensive sexuality education is different from, and pedagogically superior to, what is commonly called "sex education." The former involves education in a broader host of issues than sex, reproduction, and sexually transmitted diseases.

Comprehensive sexuality education covers all schooling years, offering material appropriate and relevant to the different class levels. It is a holistic curriculum that covers matters surrounding human bodies and sexual feelings, as well as sexual behavior that only results in pregnancy if pregnancy is wanted. Students acquire information surrounding matters of abstinence, contraception, sexual safety, and consent. Beyond these basics, however, comprehensive sexuality also educates young people about the fundamentals of gender, body image, human rights, and sexual orientation—all such issues discussed with the intent to cultivate values of respect, nonviolence, self-knowledge, and awareness. A holistic approach to sexuality education can also include issues of scale—teaching how individual reproductive decisions scale up into national and international demographic dynamics. (I am always surprised in my classes by how vanishingly little college students know about global population issues.)

In sum, the role of such a comprehensive curriculum is not only to teach correct information about sex and its consequences, but to enhance critical thinking about gender issues, gay rights, childbearing decisions, and ecological contexts. Unsurprisingly, such seemingly disparate topics turn out to be interconnected. For example, one study found that classes that emphasized "critical thinking on gender and power relationships were more than five times more effective in reducing unintended pregnancy and sexually transmitted infections as classes that failed to address gender or power."[59] The rewards of holistic sexuality education are so striking, they are worth citing at length:

Exposure to sexuality education decreases high-risk sexual behavior, delays sexual initiation, reduces the number of sexual partners, and increases the use of condoms and other forms of contraception. Studies indicate that young people who adopt more egalitarian attitudes about gender roles or who form more equal intimate relationships are more likely than their peers to delay sexual debut, use condoms, and practice contraception. They also have lower rates of sexually transmitted infections, HIV, and unintended pregnancy and are less likely to be in relationships characterized by violence.[60]

As such benefits clearly exhibit, not only does comprehensive sexuality education teach young people about human rights, it is itself a human right. All people have the right to clarity about sexuality issues; they are neither born with such clarity nor can they acquire it from their peers or families. Reliable knowledge, critical thinking, and learning the virtue of care on all matters surrounding sexuality enable the attainment of lucidity about these matters. In light of the embedded link between holistic sexuality education and human rights, it is not surprising that where opposition to such education is entrenched (for example, in some Central American and African countries), teenage pregnancy and HIV prevalence are higher than average and "girls are far less likely than boys to go to school at all, and child marriage rates are among the highest in the world."[61] Conversely, in places where comprehensive sexuality education is most developed—for example, in the Netherlands' brilliantly titled curriculum Long Live Love—rates of teen pregnancy, abortion, and sexually transmitted disease run among the lowest in the world.[62]

Worldwide, 40 percent of pregnancies are unintended.[63] (In some developed world countries, the rate is higher.) Holistic sexuality education—by raising consciousness about sexual matters along with providing sound information and tools—tends to reduce unintended pregnancies and also reduces teenage pregnancies (which are usually unintended). Among the most intriguing demographic discoveries is the finding that were unintended pregnancies minimized to the degree humanly possible, population growth would markedly slow down. (And the demand for abortions would be all but eliminated, which should inspire enlightened religious leaders and institutions to enthusiastically support comprehensive sexuality education alongside family planning and contraceptive use.) The reason that population slows down when unintended pregnancies are minimized is that some of those prevented pregnancies will occur later (when wanted), while a certain percentage of them will never occur at all; on both counts, population growth decelerates.[64]

Bringing together, on a global scale, the three human rights developments of state-of-the-art voluntary family planning, full gender equality, and comprehensive sexuality education constitutes the means —the historically and empirically proven means—for achieving the celebrated "demographic transition" to smaller families. Demographic transition, in other words, is not something that automatically happens by the invisible hand of modernity—it is something that the global community actively achieves. An international concerted effort focused on these three quality-of-life-enhancing developments would constitute a broad-tent initiative toward stabilizing and gradually reducing the global population. Such an effort would "undermine even population momentum and produce a turn-around in population growth— with the significant social and environmental benefits such a dynamic would offer—earlier than most . . . believe likely or even possible."[65]

The longer humanity waits to actively pursue these goals—universal family planning, gender equality, and holistic sexuality education— the more deeply the world will plunge into the ecological (and likely harsh social) costs of overpopulation. If many fertile women additionally embrace the option of refraining from or minimizing reproduction, so as to help alleviate the world's most pressing ecological and social problems and leave a beautiful world for posterity, the average fertility rate would shrink even further. For example, today's women and their partners might voluntarily opt for one child. Instead of climbing toward more than ten billion, the world's population would begin a slow descent within this century.[66] Moreover, as Alan Weisman pointed out in his work *Countdown*, "in the event that humanity were to agree that on an overcrowded globe we have entered a time calling for reproductive restraint, adoption is an alternative for families that choose to embrace as many children as their households can hold."[67] Expanding on this suggestion, critical theorist Donna Haraway urges, "Make Kin, Not Babies!"[68]

Were the next few generations of childbearing women to embrace this mandate for the sake of a living planet and quality of life, or even survival, of future people, how could this possibly be construed as a sacrifice or as a violation of reproductive rights? It is intelligent and compassionate action that many people might be willing to take, especially if they became knowledgeable about the planetary emergency we are in. According to the calculations of David Pimentel and his colleagues, if couples today embraced the goal of an average of one child, global population would slowly decline toward two billion in the span of one hundred years.[69] As for those who fear "soft coercion" lurking in such a proposal, they have to bypass a fact that population experts are well aware of: Some of the grossest violations of human rights are perpetrated in societies that force women to start having children when they are barely beyond childhood themselves, and to continue reproducing until their bodies are no longer fertile. (In Niger, for example, which has the highest fertility rate in the world, parents often betroth daughters before they begin to menstruate.[70])

When women achieve more educational and work opportunities, and family-planning services and sound information about fertility become available, fertility declines ensue. Societies that attain these desirable goals tend to move through a period of demographic imbalance, where the elderly population becomes large by comparison to the active workforce. This trend is already visible in some developed countries. As many observers have noted, this can present challenges for public pensions and health-care programs.[71] Researchers argue, however, that such challenges are tractable,[72] and when problems do arise, they are best faced directly instead of temporarily postponed by pronatalist policies (that is, by governments incentivizing fertility increases). As Paul and Anne Ehrlich put it, "unless you are foolish enough to believe the human population can grow forever, it is obvious that sooner or later we will have to face the consequences of changing age structures."[73]

Each country will need to address challenges accompanying an aging population, according to its own economic, social, and cultural circumstances. General approaches include encouraging higher savings rates, extending the retirement age, and/or raising taxes to support programs for the elderly. It is also crucial to shift to food policies and ultimately to a food system that enables longer, healthier, and more productive lives, while deflating bloated health-care costs tied to diet-related diseases that are preventable. Finally, while there is much media hand-wringing about aging populations in below-replacement countries, the benefits are rarely mentioned alongside the downsides. Benefits include fewer resources needed to support and educate children (since children are fewer) and less pressure to expand and maintain infrastructure.[74]

Additionally, in a century where substantial movements of people are all but certain because of climate change and ecological degradation, as well as economic dislocation and conflict, shrinking populations in developed nations can encourage greater tolerance toward immigration allowances and thus less political conflict around this issue. At the same time, prioritizing strategies for slowing and reversing population growth in developing countries will yield economic and environmental dividends that are bound to counter the pressure or desire to emigrate. And as Tel Aviv University public policy professor Alon Tal astutely observes, for "areas given to historic ethnic and geopolitical violence, family planning is nothing less than a strategic security imperative, while rapidly growing populations remain a recipe for increased enmity and violence."[75]

I deliberately exclude the approach of heightening immigration restriction as a policy for stabilizing or reducing a nation's population. In a global world, overpopulation cannot be solved as a national issue but requires international unity. Indeed, immigration debates have arguably retarded the turn to ecological sustainability by contributing to the silence (in the last two decades) surrounding overpopulation: many scientists, writers, environmental nongovernmental organizations, and policy makers prefer to avoid the population question altogether, rather than risk being associated with immigration-restriction positions that are viewed by many (wrongly or rightly) as racist. In the hullabaloo surrounding immigration matters, the much-needed international conversation about stabilizing and reducing the global population has been gingerly sidestepped.

What's more, urging the restriction of immigrants from developing nations into the developed world, in the name of an ecological cause, displays inconsistent reasoning: affluent countries cannot export environmental destruction as they do—in spades—and, at the same time, refuse entry to foreigners on the grounds of protecting their own environments from the destructive effects of overpopulation. The current global economy, founded on export-import relations, engenders what Val Plumwood called "shadow places," namely, "all those places that produce or are affected by the commodities you consume, places consumers do not know about, don't want to know about, and in a commodity regime don't even need to know about or take responsibility for."[76] It is my position that the United States, Canada, Australia, and Western Europe have rescinded any rightful claim to exclude entry to people whose lives those rich nations have actively and knowingly contributed to destroying through a self-serving commodity regime, a fossil-fuel based economy, and policies of dumping their toxic wastes in developing nations (in Africa, for example). On the contrary, taking responsibility for the shadow places that consumerism produces and requires demands that we all assume responsibility for the people who will either flee devastated places or seek to partake in the (ostensible) privileges of consumer lands.

The restless, massive movement of poor people today is driven by economic suffering, ecological degradation, and food insecurity, which are often causally tied to the trade preferences and consumer tastes of rich (and becoming-rich) countries. There are over fifty million displaced people currently in the world—globally the highest level since

World War II.[77] Moreover, the number of protracted crises from war, conflict, and political instability has grown. Indeed, "protracted crises have become the new norm, while acute, short-term crises are now the exception. . . . More crises are considered protracted today than in the past."[78] In a time of malevolent and unpredictable synergies between climatic catastrophes, land deterioration, and escalating conflicts, we can expect the numbers of refugees (human and nonhuman) and of protracted crises to burgeon. In fact, "right now, the Earth is full of refugees, human and not, without refuge."[79] Without the compass of compassion for those who are experiencing and will experience devastation as the human-supremacist plunder of the planet spikes, our global situation will become unbearably terrifying.

Overpopulation is a global problem, which is (as I hope I have successfully argued) eminently solvable by means of human-rights approaches: instituting family-planning programs, agitating for the swift achievement of full gender equality, and introducing holistic sexuality education everywhere. The international community must energetically pursue and realize these rights, while isolating as rogue pariahs the forces of "endarkenment"[80] that oppose and violate them.

Food as Relationship

Bringing humanity's population down to roughly two billion will not be the magic bullet to solve every ecological and social problem. We can rest assured, however, that it is *one* magic bullet for doing so. Significantly lowering our numbers facilitates a more harmonious way of life on Earth in at least two ways. First, that many problems—from traffic jams to health-care budgets and from garbage heaps to climate disruption—become more tractable when the dimension that magnifies them is curtailed. As Weisman states, "every emergency on Earth is now either related or aggravated by the presence of more people than conditions can bear."[81] When we consider additionally that populations are growing fastest in places that will be most vulnerable to the consequences of climate change, it becomes evident that lowering our numbers will help downscale or avoid grave harms.[82] Yet it is not only a matter of preempting emergencies but of markedly enhancing the quality of everyday life. For example, there is a yawning difference between a world of one billion vehicles (causing damage enough) versus a world of three to four billion vehicles (the direction we are headed).

There is also a huge disparity between urban areas balanced and beautified by an abundance of open, green spaces versus the all-familiar nightmare of endless road, housing, and strip-mall construction to accommodate population growth and consumer sprawl.

The second way in which significantly lowering our global population supports the turn to beautiful inhabitation of the biosphere involves, as I have been arguing, food production: A lower population will allow the radical transformation of an industrial food system that is bludgeoning ecologies, wild and domestic animals, and human well-being. The whole world can be fed: with organically grown, diversified, nutritious food; by prioritizing local and regional food economies; without mining, polluting, and dispersing the soil, but by caring for it and building it up; by abolishing CAFOs and returning farm animals to living on the land; through smaller-scale farm operations modeled on natural ecosystems; in creative interfaces with wild nature called "farming with the wild" or "rewilding agriculture";[83] and by forsaking high quantities of animal foods, for the rare consumption of such foods produced with due consideration to ethical and nutritional values. This wholesome revamping of the food system only becomes possible if our global numbers are far lower than today's. The notion that advocating such limitations carries a misanthropic subtext is hopelessly superficial if not plain daft. There is no love of the natural world without love for human beings who are, after all, created from its very fabric; and there is only phony love for human beings wherever love for the natural world is deemed as some kind of superfluous romanticism.

Food is the hub from which all spokes of life radiate. Food connects us to our bodies, to one another, to all the animals, to the land and freshwater ecologies, and to the ocean. It connects us with existence itself and with quality of being. Amidst an ecological crisis of unfathomable proportions, it is hardly a coincidence that food production is the main direct culprit. There is virtually no ecological problem that is not tied to the food system: the pollinator crisis, the freshwater biodiversity crisis, the amphibian crisis, the bushmeat crisis, and the marine life crisis, as well as excessive land use, multiplying dead zones, and rapid climate change. How can food production and trade have this scope of impact? For a straightforward reason: the current food system pretends that food has nothing to do with relationships. Because of this pretense, the dominant food system betrays all the relationships that are inherent in food: to one's body; to one another; to the wild and domestic animals; to the land, soil, rivers, lakes, and wetlands; to the ocean;

and even to the seasons. The industrial food model at best dishonors, and mostly abuses, every relationship that food entails.

An ecological civilization (a federation of bioregions as I will argue in the next chapter) will place paramount, even obsessive focus on how food is made and traded. The heart of food is relationship, and therefore how an ecological humanity understands and practices food is all about honoring relationships.[84] When the relationships inherent in food are honored, the food produced and shared will be good. When the relationship with one's own body is honored, the food eaten will be good. When the relationship with the farmer is honored—and she is paid the right wage and placed on the pedestal she deserves—the food will be good. When the relationship with the land and its waters is honored, food will be produced working with nature and all its beings, following nature's design, and at the appropriate scale. When the relationship with the soil is honored, the soil will be respected as Earth's living tissue, and farmers will work on building it in cooperation with nature and nonhumans. The dark age of pesticides will come to an end, and fertilizers will be sourced from manure and composts, not fossil fuels.

Good agriculture, wrote pioneer organic farmer Masanobu Fukuoka, "joins animals, crops, and human beings into one body."[85] One body means interrelationship. When the relationship with domestic animals is honored, they will be treated as partners—their manure is good fertilizer, and their work should be as enjoyable for them as it is serviceable for people. Farm animals must be respected for who they are: every animal is an individual who enjoys being treated as an individual and even enjoys being acknowledged by name. For people conditioned into rigid human supremacy, such notions sound soppy. Yet the truth is that every animal is an individual, and when also a member of a human household, more likely than not, enjoys hearing their name spoken. Indigenous people knew that individuality is among the core features of the natural world. Modern science is increasingly finding out.

When our relationship with wild animals is honored, they will no longer be killed to protect "livestock" and their homelands will no longer be seized—especially in order to make corn syrup, palm oil, ethanol and biodiesel, or feed for confined farm animals and fish. When humanity's relationship with the ocean is honored, fish will be taken with a light touch by artisanal fisher-folk, with due care for respecting the abundance of marine life and the intricate food webs and relationships therein. The dark age of industrial fishing will be over, and trawling will at long last be abolished. (About half the world's fish catch comes from trawling, which is among the most mindless assaults on nature ever devised. The produced silt clouds that billow underwater show up as images on satellites orbiting the Earth.[86]) Continental shelves and seamounts will be allowed their slow recovery back to life-created and life-abounding habitats.

When relationship with one another is honored, the arts of cooking and preparing nutritious food for children, families, friends, and (in the case of eateries) customers will be as well. The dark age of fast food and junk sodas will end. The quality time of slow food will spread, as human beings take the time needed to make the food, share it, and relish it together.

When we take heed of the great span of relationships entailed in human food, and how honoring these relationships changes things, we can see that food is indeed the very crux of the present predicament. As Wendell Berry puts it, with his customary simple precision, "when we change the way we grow our food, we change our food, we change society, we change our values."[87] When humanity chooses to acknowledge and honor all the relationships that inhere in food, *everything* will change—indeed everything will fall in place. Oddly enough, it is (almost) that simple.

Humanity has a different choice than continuing growing its numbers and extending its "carrying capacity" by augmenting and intensifying industrial food production. Instead, we can stabilize and gradually lower our numbers and move toward deindustrializing agriculture and fishing. It is neither progress nor necessity to occupy the biosphere with a maximum human population supported by a destructive and dissociated food system. What we can opt for at this historical juncture is scaling down and pulling back, a two-pronged approach toward a way of life that is integral with the biosphere. Besides shifts in population size and food production, other changes in socioeconomic organization will follow, especially relocalizing economies, reducing consumption and waste, and substantially downsizing global trade. These choices, prudent also for humanity's future, must flow from the desire to inhabit and thrive alongside life's plenitude, rather than plunging toward a planet repurposed and managed as a human colony.

NINE

Restoring Abundant Earth

Listen to the songs of Eden, page after page after page.   GREGORY COLBERT

Life's epic story has unfolded, without interruption, for roughly 3.8 billion years—a quarter of the age of the universe. While life's manifestations are unevenly distributed, life is a planetwide phenomenon. Some scientists have conjectured that if life exists elsewhere in the universe in the enduring and evolving form that it does on Earth, it is likely also to be planetary in scope. With the exception of the beginning and end of life's tenure on a planet, it may be as impossible to have a planet sparsely occupied by life as it is to have half an animal, as James Lovelock has memorably put it. Connections between Earth's ecological formations are as seamless as the life-forms of its sundry biomes are diverse.

Nature abhors a fragmented landscape. In the governing norm of life (that is, barring man-made fragmentation or cataclysmic events from without or within), life's natural history says this: there is always space for overflow. This is why life diversifies and fills the planet, and why no one knows exactly how many species there are on planet Earth—let alone how many there were at the dawn of the Holocene. Life writes itself in the language of freedom. In nature there is movement, tremendous fluidity enabled by the presence of habitat contiguity. Vast herds of ungulates thundering across landscapes compose moving ecological communities. Forests travel, tracking changing climates. Individual wolves, wolverines, big cats, bears, elk, great white sharks, and other large animals set off from their home base in search of mates and food and new destinations, eventually establishing new populations ("the founder effect").

With the passage of time and the interleaving of living processes, eventually mind-blowing phenomena emerge: like red knots (a bird the size of a robin) making nine-thousand-mile roundtrips, monarch butterflies accomplishing intergenerational migrations, old hub trees in forests nurturing mycorrhizal networks and smaller trees, and whale songs filling the ocean. Nature hums with fullness at every level. "Having a tendency to come together," wrote Plato in the *Timaeus*, "compresses everything and will not allow any place to be left void."[1]

The Worldview of Abundance

The composition of life-forms and their assemblies are always in a state of flux, but biochemical capacities, laws of development, processes of evolution and coevolution, the emergence of food webs, the biological construction of multispecies habitats, and the life-driven knitting of biogeochemical cycles comprise the (relatively) permanent face of life's (relative) impermanence. Earth overflows with life and its unending physical and sensory manifestations. While patchiness is ubiquitous in the natural world,[2] the biosphere tends toward a plenum.

Life resembles a kaleidoscope coruscating in slow motion. It is always the same river into which you never step twice. No place is left void, as life's plenum both fills and creates habitats: this extraordinary quality of life has no finiteness. Diversity surges at the levels of genes, varieties, individuals, species, subspecies, ecosystems, ecotones, and biomes. Sensory niches are also filled and created, enabling the expression of diverse types of awareness. *Umwelts* come into being. In the tropical forests, so thick with biodiversity, different critters occupy discrete sound-niches, ensuring that every kind of organism has its own bandwidth for expression and communication inside the vibrant geophony-cum-biophony orchestras of place.[3] Rich soundscapes once sprung within every wild habitat—terrestrial and marine—so that each locale and region generated its own unique biophonic signature, especially via the nested unique sonic niches of animals.[4] This numinous aspect of biodiversity has been silenced in our time, through the diminishment and destruction of so many of life's threads. In the ocean, the communications of whales and other beings are drowned out by the cacophony of ship engines and other mechanical sounds, interfering with marine creatures' ability to communicate, navigate, find food, and avoid predators.

To be sure, everything is constantly changing in the natural world. But there are three important qualifications of nature's constant production of novelty: one, that much of that novelty is built upon, and retaining of, what already exists; two, that typically the plenum of biodiversity becomes fuller over geological time, so that the arrow of change points in the direction of burgeoning biological richness; and three, that things tend to change slowly on nature's clock. So next time someone claims that humanity is another natural force "just changing the world," it would serve to remember that such casuistry rests on ignoring how very differently nature changes the world.

Change in nature supervenes from the evolved natural order within which change flows. Thus life tends to retain established anatomical forms, laws of development, biochemical patterns, and organismal alliances, while building—excruciatingly slowly from the perspective of human life—new life-forms, patterns, and relations. On the surface, the natural world is ever in flux, but foundational aspects of life persist, forming the enduring frameworks that undergird transformation. Most especially, contiguity of ecological configurations reigns; complexity of ecosystems ensues; diversity of species, subspecies, and populations mounts; life's molding of physical and chemical environments prevails; biochemical associations between diverse organismal groups become enduringly interlocked; and evolution—which preserves sameness-undergoing-change in response to environmental shifts and pressures—is the perennial name of the game.

Most things in nature also tend to change slowly: there is time for life to adjust, time to move, time to evolve. Only with catastrophic events do changes occur abruptly, and if the catastrophes are global, so are the consequences: with asteroid strikes, flood basalt protracted eruptions, or a global culture stripping the natural world of anything that can be called a resource, there is no time for life to adjust, move, or evolve. Thus, when we look and see disturbance all around us, and find ourselves primed for a slanted interpretation of natural history as intrinsically characterized by "disturbance," we might take pause before assigning disturbance so facilely to nature's foundational makeup.[5] To paraphrase poet David Ignatow, "we should be content to look at a mountain for what it is and not as a comment on our life."[6]

Earth's story is life's story, whose phenomena emerge in each place uniquely and over the whole planet diversely, always entangled at local, regional, and global levels. Life occupies niches and proceeds to construct them by changing the geomorphology and/or biochemistry of its surroundings.[7] Different life-forms accommodate other life-forms by means of such niche building, and also by their edible, breathable, or otherwise consumable waste byproducts, as well as by their eggs, larvae, offspring, and corpses becoming nourishment for others. With the exception of mass extinction events, life is always enabling more of itself to surge. There is ceaseless mutual feeding on one another and on each other's byproducts, as well as the cocreation of a physical and chemical environment in which more life is supported to flourish.

The molding of physical and chemical environment becomes increasingly significant the greater the proliferation of organisms.[8] Consider the example of the gopher tortoises of the Southeastern US longleaf pine ecosystem where, for two million years and until recently, they excavated vast networks of burrows that supported hundreds of species of reptiles, mammals, birds, amphibians, and insects.[9] "Only two-thirds of a longleaf forest ecosystem is visible," reports Tony Hiss of the *Smithsonian*, "with the rest underground." "Three hundred and sixty animal species take shelter in the 40-foot-long, 10-foot-deep burrows excavated by shy and dusty gopher tortoises," he continues. "They retreat down these paths to where fires and hurricanes can't penetrate, and where temperatures never sink below 55 degrees in winter or get above 80 in summer."[10]

By spreading plant seeds, and by their burrows engineering the underground hydrology, gopher tortoises also shaped the plant life above. By virtue of designing the ground's architecture, they invited countless beings to move in, and through their collective participation, the foundation of all there is—soil—was made and remade. The resilience of such strength-in-unity ensembles of life is evident in their longevity. Two million years is a timeline one order of magnitude older than our species' existence; it is also a timeline that saw numerous climate shifts and included Native American inhabitation and fire use in the region. The primary habitat of the tortoise was the longleaf pine forests, which European settlers reduced by 97 percent, leading to the decimation of the tortoise population by 80 percent.[11] Gopher tortoises continue to be in steep decline from human population growth and continued development.[12]

The point of this example is to underscore that human expansionism—beyond diminishing habitats (like the longleaf pine forests) and devastating keystone species (like the gopher tortoise)—crushes the creative force of diverse, abundant, and enduring life through which life-rich topographies are generated and sustained. The point is also to highlight that the impoverishment human expansionism causes runs contrary to nature's order. The world that gopher tortoises create—or beavers, bison, fungi, mosses, hub trees, coral reefs, sea otters, elephants, and sharks for that matter—invites us to pause and view it. In viewing their world, we might understand it as the imprint of a "worldview"—call it the gopher tortoise worldview—on the landscape. While the human supremacist worldview constructs a world of low species diversity, homogeneous ecologies, and humanized landscapes, the worldview of gopher tortoises shapes a world that accommodates a great diversity of beings and an enduring web of life. Whose worldview is more equitable, beautiful, and resilient is hardly a competition.

The story of gopher tortoises cocreating the land and nurturing a community of diverse life is the story of life everywhere; only the specifics differ. It is always the story, in eco-philosopher Freya Mathews's words, "of symphonic synergies in which the elements of nature intimately shape one another and collectively achieve the great metabolic processes of Earth."[13] If it is near impossible to imagine that there may have been as many as ten million whales living at once in historical times,[14] it is even more difficult to picture the abundance of life in the seas that sustained them *and* billions of marine predators. Writers of centuries past "marveled at the fecundity of the seas," and struggled to find words to describe it. The vast empires of herring, with which I began this work, are a case in point. One eighteenth-century naturalist stated that they were "composed of numbers that if all the horses of the world were to be loaded with herring, they would not carry the thousandth part away." Another wrote that "when the main body arrived, its breadth and depth is such as to alter the appearance of the ocean." "Herring along the coast are so tremendous," a Danish chronicler reported, "that they block the passage of boats and could be taken by hand." And where the herring went followed those who ate them—lavishly numerous and diverse—the whales, sharks, tuna, swordfish, cod, seals, and seabirds, among others.[15]

Every tier of marine life, from the invisible sun-shaped diatoms to the majestic blue whales, abounded: the anchovetas off the coasts of South America, the herring of Europe, the menhaden of eastern North America, the sea turtles (and big fish, manatees, monk seals, and crocodiles) of the Caribbean, the billions of Atlantic cod, the sturgeon of Chesapeake Bay, the pelagic sharks of the Gulf of Mexico, the oyster beds of Dogger Bay, the totoaba and groupers of the Gulf of California, the adaptive radiation of rockfish from Mexico to Alaska, the prolific jack mackerel of the South Pacific, and the cosmopolitan citizenry of hundreds of millions of whales, dolphins, large sea turtles, tuna, swordfish, and sharks, among others.

Marine creatures forged the ocean's morphology with their bodies, binding the seabed as the oysters did, adding multidimensionality to habitats as the reefs, sea fans, mussels, and others did, and raining their miniscule shell-homes down as billions of single-celled algae have effected, whose tiny calciferous houses, over the long haul, submerge under tectonic plates and reemerge in the atmosphere, in the form of carbon dioxide, through volcanic activity.[16] Yet life does not only contribute to building the physics and chemistry of the world. It knits and mixes the world's components. The abundance of marine beings were one of the tremendous forces that churned the ocean, helping to mix the waters vertically and horizontally. In the words of writer and diver Julia Whitty, "life itself helped to rotate the great waterwheel of temperature and salinity driving the underwater rivers of the world and making our planet habitable."[17]

The ecological baseline of the ocean's extraordinary fecundity speaks to the nature of life itself, to its celebratory timbre and the inherent absence of scarcity on a life-shaped planet. Life's entanglements also tell us about the feast of the senses and the magnetic pull of different forms of awareness toward one another. Consider this: Any sizeable object floating in the midst of the ocean draws a myriad of creatures to it. Why? "Nobody knows," writes marine conservation biologist Callum Roberts, "exactly why fish gather around floating objects in the open sea."[18] If they come to eat—don't they also risk being eaten? They come because life is aware and irresistibly attracted to other life and to the lure of mutual witnessing. Indeed, mutual witnessing can take front seat to feeding, which is why expert divers are able to swim with large and potentially dangerous sharks as long as eye contact is maintained—as long as all parties are aware that everyone is aware.[19]

Undoubtedly, "removing a few million whales, a few million sea turtles, and billions of long-lived fishes"[20] has had a profound effect on every aspect of the ocean (biological, physical, and chemical), while simultaneously impoverishing the human experience and understanding of Earth's magnificence. The mass extermination of marine life and its habitats—effected especially by the rapacity of industrial fishing, but also by massive contamination, climate change, and acidification—is not only driving extinctions and ecological unraveling: it is extinguishing life's ability to transform and to cocreate both the world and more of itself. By decimating living beings and their homes, we vandalize the creativity of life. What's more, for all we know, by fishing out the seas, undermining the base of the ocean food web, and disabling the ocean layers from mixing, we may be threatening to overturn fundamental, long-term biochemical parameters of the planet, including oxygen levels in the ocean.[21] And by recklessly perturbing the sleeping dogs of the ocean's methane deposits through increasing Earth's average temperature, we risk handing over the reins of mass extinction to rapid climate change—an eventuality that would amount to "endgame" for all intents and purposes.

Nature abhors a fragmented landscape: even at the borders between different ecologies, novel life assemblages spring forth. The place where distinct natural communities converge, ecologists call *ecotone*. The ecotone forms a wide boundary where different ecologies overlap and mesh into one another, preserving elements of both while giving rise to new living phenomena and configurations. Some fabulous mixes happen in the volatile thresholds where land and sea meet. In the world's remaining coastal rain forests, for example, "there is no graceful interval between ocean and trees. . . . The boundary between the two is unstable, and the sea will heave stones, logs, and even itself into the woods at every opportunity." There, the forest floor and canopy are "almost literally seething with life." Salmon and trout might get stranded on branches, murrelets (birds) fly underwater, seals pursue fish deep into the forest, and "the patient observer will find that trees are fed by salmon, eagles can swim, and killer whales will heave themselves into the graveled shallows and stare you in the eye."[22] Free rivers teemed with fish even recently, and rivers of fish swimming inland from the ocean might be pursued by predators, giving rise to ecotone phenomena: porpoises could be sighted in Britain's Thames River and bull sharks in North America's James River.

In the unbounded space where ocean meets terra—when that space is left free to be self-arising—the cocreation of life and inorganic environment, along with patterns of spontaneity, adaptation, seasonal fluxes, stochasticity, nutrient cycles, biological fecundity, and nonhuman awareness all mingle into a fantastic tapestry that resembles an outburst of wild Earth's genius and imagination.

A popular cliché has it that life is all about struggle, competition, and selfishness. Let us call this perspective out for the babble that it is: for only within a life-filled planet, a *lifeworld*, do phenomena of struggle, competition, and selfishness arise and pass away in their relevant contexts. Suffering happens, as does violent death, disease, and extinction: these are all part of life but not its foundation. "Life," biologist Lynn Margulis wrote, "is a network of cross-kingdom alliances that help keep the entire planetary surface brimming with life."[23] In its essence, life is about coexistence, interdependence, abundance, diversity, and endlessly varied relationships. Life is "the universe breathing beauty."[24] Life is an extravagant, self-sustaining oasis that defies—for as long as it is able—the decline and dispersion into disorder known as entropy. The vibrant lifeworld is all-encompassing in the phenomena it manifests and cannot be reduced to a one-dimensional ideological picture, be that picture malevolent and ruthless or benign and safe. Yet there is the one thing we know beyond any doubt: that the lifeworld is all good. Do we not see, as poet Rumi exclaimed long ago, that we have fallen into a place where everything is music?[25]

The nature of life is interdependent entanglement. The natural world is blended as a commonwealth—the wealth of Earth inhabited in common by all life. "There is no possibility of a detached, self-contained local existence," wrote philosopher Alfred North Whitehead, "the environment enters into the nature of each thing."[26] Each living thing in turn, not to mention their collective power, enters into and shapes the environment. The biosphere is co-fashioned as life's home by the totality of life. Ninety-nine percent of the air is made by life inhaling and exhaling. Life in the ocean makes 50 percent of the atmosphere's oxygen. As marine biologist Boris Worm is fond of saying, you can thank the ocean for every other breath you take. The tough root systems of trees and other vegetation carve the course of rivers; indeed, the rise of trees in the Carboniferous (around 300–350 million years ago) played a star role in bringing deep, branching rivers into being.[27] Freshwater life fills and snakes along the paths that rivers course, while terrestrial beings team their banks.

Life is relationship, and relationship, as indigenous cultures have long celebrated, is about reciprocity. As Margulis quipped, on Earth "there are no virtuoso individualists."[28] Through organism-mediated processes, the land brings nutrients to the seas, and the seas, also through organism-mediated processes, return nutrients to the land. Beings in the seas' upper layers sustain the strange menagerie of abyssal creatures, and organism-created nutrients in the depths well up and nourish beings in the upper zones. The creatures living in tree canopies never meet the creatures living in the soil: never mind, for the two are deeply associated, with life-filled canopies feeding critters in the understory, and life in the understory feeding the trees and all who live in their sun- and cloud-kissed heights.[29] All forest trees, including different species of trees, connect and nurture one another underground.[30]

Trees and fungi are fast friends: tree roots deliver sugars to the fungi, and fungi give trace minerals, through the roots, to the trees. Ancient forests keep rivers and streams cool, banks stabilized, and running water sediment free: this supports migrating fish to flourish, who, in turn, feed the entire forest. Plants offer berries and other fruits to wild animals, who then spread the plants' seeds inside "fertilizer packets." "The fate of many plant species," writes indigenous author and botanist Robin Kimmerer, "lies in the hands of complex interactions with pollinators, butterflies, bees, bats, and hummingbirds."[31] Large herbivores accelerate the nutrient cycle of ecosystems through consumption and defecation, returning nutrients to the land orders of magnitude faster than processes of leaf loss and decay alone.[32]

A suite of life-forms from all five kingdoms of life are involved in soil building, life's living ground. For any location, "soil is made up of the flesh and bones of every creature who has passed this way in the last few millennia."[33] Spongy mosses protect the soil, capture and time-release moisture, and create microecological universes for germinating plants, insects, and other life. "Rivers give thanks to mosses," writes Kimmerer, for "without the mossy forest to hold it, the water runs brown with soil, silting up salmon streams as it carries the land to the sea."[34] "What is good for mosses," she adds further down, "is also good for salamanders, waterbears, and wood thrushes."[35] Trees also contribute to protecting and creating the soil, as the litter created by fallen leaves and canopy animal debris slowly decomposes. When trees die and fall over, they return the nutrients of their bodies to the landscape while also creating habitat for more life. Earthworms, insects, and other animals mix the ground organics into humus. The fungi and bacteria break it down even more. Water is vital to all these processes; indeed, some forests are rainmakers while others collect fog and distill its watery flow into the forest. Microorganisms of the ocean and atmosphere also do critical work in precipitating rain.

From life's choreography—where phenomena of physics, biology, biochemistry, behavior, awareness, communication, and chaos jostle in established and spontaneous patterns—great proliferations are created. Vast flocks of birds adorned skies, wetlands, and seashores worldwide. Abundant land, sea, and air animal migrations have not only told the seasons' stories, but contributed to bringing the seasons into being. Hundreds of millions of eggs pile over vast areas of the seabed, or wash to the sea's edge, feeding multitudes of creatures before a fraction develop into the organisms that spawned them. The explosion of horseshoe crab eggs on the shores, intertidal zone, and shallows of the

Delaware Bay were a vital food link for the long-distance traveler, the red knot; the horseshoe crab eggs sustained their now dwindling migration.[36] The eggs and larvae of carnivorous fish, such as cod, feed fish who will later become the prey of the cod who survive into adulthood; mutual feeding of prey and predators is thus a literal and common phenomenon in the ocean.[37] Prey species proliferate wildly in response to the pressure of their predators—there were incalculable numbers of marine creatures to sustain hundreds of millions of sharks and other marine animals who graced the planet.

Enormous herds of ever-on-the-move herbivores do not decimate the grasslands, but on the contrary the grasses grow because of them, and the animals and grasses (along with other life-forms) together build more soil. In East Africa, for example, a grazing succession of zebras, wildebeest, and Thomson's gazelles create vegetation conditions that benefit one another in succession, and together these traveling grazers "increase the vegetative productivity of the grasslands."[38] In the North American Plains, prairie dogs, bison, and pronghorn antelopes similarly benefited one another and the land they shaped and inhabited.[39] Wherever we look in the natural world, we find what Kimmerer calls "the covenant of reciprocity."[40]

The intermingled manifestations of life on Earth, when Earth is left free to manifest them, are breathtaking. Tibetan people have a lively expression for the human outburst of amazement mingled with gratitude: *Emaho*! This place is our place in the universe—this beauty, goodness, reciprocity, and wealth of life's overflow with endless ways of knowing and being contained therein. Humanity does not need to invent a new worldview to live by, for life's worldview of abundance created who we are and continues to envelop us. That worldview is deep within us, though "it remains buried," in the words of environmental thinker Paul Kingsnorth, "in the ancient woodland floor of human evolution."[41] We have the choice to recognize that worldview, let it manifest and become again the cosmos it intrinsically is, and inhabit it. "When humans surrender the arrogance of domination," writes philosopher Erazim Kohák, "they can reclaim the confidence of their humanity." "Nature," he continues, can then "accept the human as also part of its moral order."[42]

Rewilding the World

Eons after life emerged, it slowly built itself up until, by virtue of its abundance, it began to contribute to crafting the atmosphere, lithosphere, and hydrosphere. Wilderness came out of this mash-up of the living and the abiotic, a synthesis greater than the sum of the two. All its dangers notwithstanding, it is highly inviting for residence. Our species as well came into being as the unique biological entity we are out of the wilderness.

Moving toward an ecological civilization begins by embracing the epic, deep truth of Earth as a living planet that creates diversity, complexity, biological wealth, and a stunning array of forms of awareness. Wilderness—or free nature by any other word—is the cauldron within which Earth performs the alchemies that create such splendor. Re-creating civilization will have us turning toward designing a thriving coexistence of human and nonhuman communities, interpenetrating human history with natural history into uncharted realms of beauty, diversity, plenitude, and freedom for all.

Thriving is the energy that underpins wild nature. It underpins the physical reality of wilderness, for within big, contiguous landscapes and seascapes the cosmic dynamics of ecology and evolution can fully unfold, and life's diversity of species, subspecies, varieties, populations, genes, and ecologies blooms. Thriving also underpins the ethics of wilderness. Embracing freedom for all beings means cultivating and preserving the conditions within which their innate makeup can self-realize, and within which they might achieve their fullest individual, ecological, and evolutionary potentials. For many wild beings, that requires large-scale stretches of nonfragmented and un-interfered-with terrestrial and marine places. As for humanity, the ethical imperative of freeing nature redirects us to live respecting appropriate boundaries within the full house of life.

The ethical ramifications for nonhumans and humans are ultimately not discrete: they intersect in the profound space we call dignity. Restoring big wilderness in the biosphere restores dignity to all beings to be and live as who they are—not as starving, persecuted, and exiled refugees teetering on an edge between living and dying, or existence and extinction. It also restores dignity to human beings, for the rampage called "civilized behavior" on Earth promotes and institutionalizes such rank attributes as greed, arbitrary exercise of power, and arrogance, and incites killing, experimenting on, harming, and enslaving beings with impunity.

The most encompassing context for the thriving of all life is sustaining the biosphere's freedom to express its inherent nature. The biosphere's nature is to create diversity of life-forms, plenitudes of wild beings, complex and dynamic ecologies, and extraordinary living phenomena such as animal migrations, biodiverse ancient forests, intricate mycorrhizal networks, and fascinating variations of intelligence and awareness in all life's kingdoms. For the biosphere to be free to express its nature, we must pledge vast areas of continents and ocean to remain unoccupied, unexploited, and connected. Being-wilderness is the original blueprint of the biosphere and the precondition to express itself "as a work of art that gives birth to itself."[43]

For rewilding the biosphere and ourselves within it, I propose three frameworks to think with that provide pragmatic and visionary directions forward. The first and most immediately achievable goal is robustly protecting 25–75 percent—or a rough and memorable half— of Earth's biomes. This is known as the platform of "Nature Needs (at least) Half," or in E. O. Wilson's wording, and recent book title, "Half Earth."[44] The second framework taps into Roderick Nash's vision of "Island Civilization," calling us to reimagine human communities integrated within the vastness of free nature—as opposed to the current inverse status quo. Finally, the third framework is the longer-term goal of designing human inhabitation on bioregional principles, which, like indigenous ways of life, invite the creation of distinct but interconnected human cultural-economic identities fashioned in reciprocity with geographical place and grounded in love for all its beings.

Protected areas today are the havens for Earth's remaining biological wealth, ecological complexity, and evolutionary potential. They must remain formally protected, until the time comes when such areas are no longer needed. Conservation practice is thus "part of a larger strategy," as Thomas Birch put it, to make all land and seas into "sacred space, and thereby to move humanity into a conscious re-inhabitation of wildness."[45] The entire Earth can then become what David Brower envisioned as "a conservation district in the universe"—Earth Park, except that the word "park" will be as unnecessary as human-nature legal boundaries.[46] Protected areas, however, are indispensable until that day when human beings create a way of life that has left in the dustbin of history all commerce in wildlife body parts, as well as such irrational trade-offs as wetlands for cane sugar; rain forests for meat and palm oil; prairies for corn and soy; intact ecosystems for coltan, diamonds, gold, or oil; mountains for coal; sagebrush landscapes for natural gas; boreal forest for butamine; and a life-filled ocean for trash dumping and mass-extermination-begotten seafood. Until an enlightened time dawns, nonhumans and their places must be shielded with strict laws, real enforcement, and militant vigilance, if they are to survive.[47]

225

To avert the impending mass extinction, the outward expansion and connection of nature's free spaces is imperative, by means of increasing the area of protected land and seas well beyond what is presently allotted—roughly 15 percent for land and 3 percent for the ocean.[48] Vast portions of land and seas must be shielded from infrastructure, crop and animal agriculture, human settlements, mining projects, industrial fishing, and shipping lanes. The intertwined emergencies of a hemorrhaging biodiversity and catastrophically changing climate have inspired bold proposals from scientific and conservation communities.

One mandate is the cessation of all primary forest destruction—boreal, temperate, and tropical.[49] The dry and wet tropical forests of the Americas, Africa, Asia, and Oceania harbor a stupendous diversity of known and unknown life-forms, and their conversion into poaching grounds, cattle ranches, soybean monocultures, oil palm plantations, and other agricultural as well as mining ventures is a clarion sign of a global economy gone haywire and, by any standard of rationality, insane. There exists a general impression (encouraged, undoubtedly, by the view of forests as "renewable resources") that ancient forests can regenerate after exploitation. This is a convenient fiction, for the biodiversity of primary forests does not simply bounce back. In former agricultural land, for example, after forests regrow (if they are allowed to regrow), their native species diversity is diminished and their species composition is characterized by "an increase of common, competitive species at the expense of ancient forest indicator species."[50] In other words, biotic homogenization prevails in postagricultural ecologies, contradicting a widespread notion that forests recover from past agriculture or other ventures.[51] As a global community, we must find the wisdom and the will to let ancient forests be: the most prestigious art museum of the world cannot hold a candle to them, unless its walls were to start breathing and its art objects were to come to life.

Another mandate is to vastly expand protected areas in the ocean, 97 percent of which is currently open to fishing and only 3 percent protected. Oceanographer Sylvia Earle calls for inverting this ratio— protecting 97 percent of the ocean.[52] Indeed, a recently discussed proposal has been to stop all fishing in the high seas.[53] Coupled with expanding protected areas along the coasts, full protection for the high seas would allow marine life to recover from the industrial plunder of the past century. Instituting this measure would signal our capacity to match our will to the knowledge of the marine devastation that continues under our watch. It would reflect our wisdom to treasure marine biodiversity for its own grandeur and for the sake of the experience and livelihood of future people.

Harking to Reed Noss and Allen Cooperrider's pioneering work in the 1990s, the initiative Nature Needs Half mandates safeguarding 25–75 percent of every biome (with the exception of primary forests that should hew toward 100 percent protected). Such large-scale nature conservation aims to protect all Earth's ecosystems, viable populations of wild animals and plants, and ecological and evolutionary processes.[54] Protected areas must be representative, sufficiently large, unfragmented, and connected. Ecological restoration is also a partner in this work, involving measures like removing dams, roads, and fences, and reintroducing extirpated species. E. O. Wilson calls such protected areas "biodiversity parks," describing them "as a new kind of park that won't let species perish" and "as the conservation of eternity."[55]

Nature requires protection at such an enormous scale in order to redress the extinction and climate crises. The urgency of expanding and connecting protected areas today cannot be overstated. Conservation efforts and funding for protected areas in the developing world—where wild animals and ecologies are in steep, rapid decline—need to be increased. (Only 10 percent of total funds go to the developing world's nature reserves.[56]) What's more, past mistakes alongside successful conservation projects have shown that local people must be integral and active partners in all nature protection endeavors.[57] Parallel to respecting the right of local communities to be a formative force in conservation is, in the words of environmental thinker Helen Kopnina, the "ethical, political, and practical call to include the rights of nonhuman actors in the discussion of environmental justice."[58]

Continental- and oceanic-scale conservation—based on cores (such as parks, wilderness areas, marine protected areas, and restored places) and linked via corridors—constitutes the only effective means to staunch biodiversity losses, stop the nonhuman genocide of extinction, and soften the blows of extreme dryness, extreme wetness, extreme heat, and other extreme weather events that come with anthropogenic climate change. In the words of conservationists Kristine Tompkins and Tom Butler, the Half Earth vision, "puts the largest possible framing on contemporary conservation efforts. Parks and wilderness areas aren't just about scenery or recreation or campfire cookouts with the family during summer vacation. They are the fundamental building blocks of a durable future—for humanity and all of the other species who call this planet home."[59] As E. O. Wilson writes, "there is no solution available . . . to save Earth's biodiversity other than the preservation of natural environments in reserves large enough to maintain populations sustainably. Only Nature can serve as the planetary ark."[60]

In substantially pulling back from incursions on terrestrial and marine nature, we do not "save the Earth," which after all has a small eternity to reinvent itself. In pulling back our presence and interference, we set the biosphere free by means of creating or finally reclaiming a deeper sense of who we are. In that covenant, all earthlings including ourselves regain the biosphere's vastness, mystery, creativity, and beauty. "We are earnest to explore and learn all things," wrote Henry Thoreau, yet we yearn that "all things be mysterious and unexplorable, that land and sea be infinitely wild, unsurveyed and unfathomed by us because unfathomable."[61] Letting Earth be its own infinity *is* the final frontier—the forever frontier for visiting, exploring, and imbibing with senses, mind, and spirit.

Environmental entreaties against limitless growth on a "finite planet" risk propagating a fiction to make an otherwise valid point against human expansionism. For Earth's proverbial "finiteness" is by no means an innate quality of its spherical corporeality, or net primary productivity, or net solar input, or any other quantitative measure. Earth's perceived finiteness and present and coming scarcities for humans and nonhumans are straightforward results of human imperialism's nonstop plunder. Expansionism has produced the illusion of finiteness, while threatening to make that illusion real—for a finite time (whatever that ends up being) of human existence. In itself, however, Earth is an infinity: an infinity waiting to be remembered, an infinity waiting to be restored, an infinity waiting to be reinhabited by a new humanity. Earth is a planet of "inexhaustible vigor and Titanic features."[62] It is that planet that the human mind, body, senses, and spirit require to thrive—regardless of what other worlds or possibilities technologies may open. Abiding by Earth's worldview of abundance is the only pathway and foundation for achieving world peace. Restoring abundance requires scaling down and pulling back the human enterprise, laying the weapons of nature colonialism down—and thus leaving behind all the social inequalities and conflicts that are corollaries of that colonialism.

Notwithstanding the platitude that wilderness restoration on a planetary scale would have humanity "going back to the caves," the recovery of a lush world and the reinvention of humanity as loving member of creation do not demand the wholesale overthrow of civilization. Nor do they call for the blanket rejection of all modern technology. Rather the emergence of an ecological civilization rests on "conceding," in the words of author Stephanie Mills, "that nature is unsurpassed in its genius at evolving niches, creatures, and adaptations beyond counting."[63] Nature's unsurpassed genius is the abiding context and ground of human advancement—it creates nourishment at every level. The human occupation of the biosphere, however, in burgeoning numbers, excessive consumptive patterns, wasteful trade, industrialized landscapes and seascapes, and unrestrained sprawl over the face of the Earth will tether an albatross around humanity's neck. If anthropogenic mass extinction is left to run its course unchecked, how will the human psyche bear that onus? We must awaken to the unthinkable burden we are passing on, and put a stop to this catastrophe now.

Roderick Nash proposed the general contours for enlightened human inhabitation within a rewilded biosphere in his idea of Island Civilization.[64] Instead of civilization being the sea within which fragments of derelict and diminished wild nature persist as fragile islands, he argued that the reverse should be our goal: wild nature can become the sea within which human habitats are integrated. In my view Nash's vision falters in the details, for it restricts people to densely populated urban settlements—a way of life too regimented and homogeneous to meet the variety of human aspirations. The main thrust of Island Civilization, however, is sound. Humanity can reside inside the space created by wild nature, rather than wild nature being pockets inside the geographical space controlled by humanity. In Island Civilization, humanity's population and economic ventures will be downscaled. Simultaneously, vast parts of the world are freed from infrastructure—roads and shipping lanes, pipelines, electrification grids, and communication towers. This core vision sees humanity inhabiting the expansive beauty and diversity of wild Earth. Every breath human beings take, no matter where, will be "Earth bathing," while the waters of all streams, rivers, and lakes will be drinkable and swimmable. Land and seas will offer a feast for the eyes, for all the senses. "Pleasure," reflected a twentieth-century mystic, "is an attribute of paradise and . . . it must be earned."[65] For enduring planetary pleasures to envelop and delight humanity, we must give up the cheap thrills of mass consumption that create hell on Earth.

The places at the borders between unbounded wilderness and human rural and urban settlements can be designed as real anthromes, in which the cultivation of the land, in partnership with nature, creates landscapes that comingle cultivated and wild elements and are uniquely biodiverse in their own right. From behemoth, life-crushing, one-size-fits-all industrial petro agriculture, we can transition to agro-ecological, locally designed, diverse, and resilient small- and medium-sized farms. In the words of agriculturalist Fred Kirschenmann, "we will need to pay much more attention to the uniqueness of each eco-logical 'neighborhood' and design agricultural systems that are suited to each ecology, rather than imagining another uniform, homogenized, global agriculture typical of [industrial] agriculture."[66] From bludgeon-ing industrial fishing, we are called to return to a level playing-field of artisanal, low-impact fishing. And from factory farming, farm animals must be reintegrated into the land where they can live good lives and contribute their gifts. Food consumption will center on an authentic "nutrition transition," where vegetables, legumes (which include beans, lentils, and peas), fruits, grains, and nuts are supplemented by animal products. The human desire for culinary variety can be fulfilled by the ingenuity of real farmers and by a food culture that draws on global cuisines, even as the long-distance transportation of food is reined in. Robust constraints in the productivity of artisanal fishing and farm ani-mal grazing mean that the consumption of animal products will be fru-gal: but what is consumed by way of fish, meat, dairy, and eggs will have a higher nutritional quality and be sustainably and ethically produced.

An Integral Way of Life

"Imagine a way of life," writes bioregional thinker Kirkpatrick Sale (paraphrasing an anthropologist's description of a Native American tribe), where people are "deeply bound together with other people and with the surrounding nonhuman forms of life in a web of being, a true community in which all creatures and all things can be felt almost as brothers and sisters and where the principle of nonexploitation, of respect and reverence for all, is as much a part of life as breathing."[67]

The third framework to think with in connection to restoring Earth's abundance is modeling human communities on the lifeway called bioregionalism. Bioregionalism is a mode of inhabitation that has been explored since the 1970s by a cadre of deep ecologists, think-ers, and activists, and it differs profoundly from the current models of the nation-state and economic globalization.[68] The integral inhabi-tation of indigenous peoples has inspired the bioregional vision of human life. "The earth and myself are of one mind," asserted Chief Joseph of the Nez Perce tribe. "The measure of the land and the mea-sure of our bodies are the same."[69] Elsewhere, he voiced the wisdom of nature preservation: "We were contented to let things remain as the Great Spirit made them."[70]

Because of the cavernous gap between the bioregional vision and present-day rapacious realities, discussion of bioregionalism runs the risk of being labeled utopian—and, like George Orwell once quipped, "all utopias are swindles."[71] Yet as sociologists have cogently argued, ideas are as fully implicated in human action as ideals are in social change and intentions are in planning.[72] Therefore, cultivating a lively conversation about a human life integral with the biosphere is anything but a swindle or a waste of time. In what follows I offer a description of the bioregional ideal "not as a manifesto but as a conceptual vocabulary, not as a doctrine to be followed but as a set of principles to be explored."[73]

Bioregionalism is the name given to the political, economic, and cultural design of communities endowed with the advanced consciousness of inhabiting Earth's abundance. In the bioregional way of life, human inhabitation will fit within the biotic and abiotic features of each bioregion. "The bioregional movement," according to activist John Davis, "seeks to recreate a widely shared sense of regional identity founded upon a renewed critical awareness of and respect for the integrity of our natural ecological communities."[74] A bioregion is a geographical location characterized by a topography, animal and plant communities, soil types, bodies of water, weather patterns and microclimates, humidity and aridity gradients, animal migrations, human histories, and other unique features. While nature's boundaries are always fluid, every bioregion has fairly stable and recurrent natural patterns that give it its peculiar sources of livelihood, vistas, lore, and feel. In allegiance with the primacy of bioregion, with its multispecies constitution and inherent beauties, political boundaries must be redrawn, in environmental writer Bron Taylor's words, so as "to reflect the natural contours of different ecosystem types."[75] According to Sale, living in intimacy with a specific locality means being "in touch with its particular soils, its waters, its winds . . . learning its ways, its capacities, its limits." He goes on to add: "We must make its rhythms our patterns, its laws our guide, its fruits our bounty."[76]

Bioregional inhabitation is all about creating cultures in partnership with place—"a poetics of fit," in the words of Deborah Bird Rose.[77] A bioregion is as much a delimited territory as it is a hybrid space that is both "a geographical terrain and a terrain of consciousness."[78] Energy production, food cultivation, housing and settlement design, material culture production, and artistic and recreational expression align with the characteristics and capacities of specific places. Bioregional economic life is grounded in sustaining—not exploiting and drawing down— the natural world. Bioregional economies will largely utilize local and regional materials and labor, conserve and recycle, and eschew pollution and waste. "Drawing a lesson from the indigenous model," writes Val Plumwood about bioregionalism, "we need to develop forms of life and production where the land of the economy (production, consumption, and service provision) and the land of attachment, including care and responsibility, are one and the same."[79] Bioregional living means no more "shadow places," *anywhere*, sacrificed for serving consumers somewhere else.

In agreement with food sovereignty advocates,[80] bioregional food economies are to be grounded in the only realistic approach to food security there is: the capacity to produce (and forage) sufficient nutritious food nearby, while respecting the autonomy and integrity of the more-than-human world, and supplementing additional food items by means of trade. Integral food systems will thus not create situations of distant food dependency nor force faraway lands to make luxury foods for the cornucopian coffers, and out-of-season eating, of the privileged.

There is no place for absentee landowners in a bioregional economy nor for a food system that kowtows to global trade. At the same time, the robust self-sufficiency praised by bioregional advocates is not a call for insularity: it follows from the fact that there is no common sense in depending on imports for basic needs nor in allowing imports and exports to define a society's lifestyle. Moreover, in prioritizing eating locally produced food, as naturalist Stephen Trombulak observes, we become "mindful of the relationships among beings—pollinators, decomposers, predators, producers—that make the food available." We give thanks "to all of [our] neighbors almost daily for the beauty, mystery, and community that together we create and in which together we live."[81]

Bioregional peoples do not sell out the places they inhabit for such dissociated goals as money and power. Bioregionalism is foremost "an ethics of loyalty to place"[82] and all its inhabitants. Such a lived ethics fosters a sense of loyalty that extends beyond any specific geographical location to encompass all Earth's beings, peoples, and places. For Earth is, after all, our bioregion in the universe.

The guiding principle of bioregional inhabitation is that there is sustenance, space, and freedom for all—human and nonhuman—to enjoy the freedom to express their natures and to move in the direction of self-realization. "Population control" will not only be eschewed but also unnecessary in the political economy of the bioregion for two interrelated reasons. One, women (like all human groups) will enjoy equality and thereby be in control of their fertility, while the bioregional polity will provide the family-planning services that are required. As a result, most women will choose to have zero, one, or two children; the latent desire for few (if any) children, discussed earlier, will become the norm, since women will have real choice along with the means to exercise it. Second, a universal moral norm shared by bioregional cultures will be that each child's life and potential are valued in themselves. The paramount goal of raising children is to provide them the conditions to pursue self-realization. Any reason that has ever existed to bring children into the world, other than their intrinsic right to be and flourish, is out of tune with a morally advanced civilization: children are not to be begotten for old-age insurance, or to labor in sweatshops or agricultural fields, or to keep the economy growing, or to aggrandize the armies of nationalistic tyrants. Nor will human beings be brought into the world so they can be given "jobs." That vulgar word should have no place in bioregional societies. (Wendell Berry once aptly likened its ring to "throwing a bone to a dog.") Meaningful work and service are the valid economic-cultural frames around which human beings build their lives, direct their destinies, and contribute to the overall well-being of their communities.

Ecological education, as author David Orr has eloquently argued, is an indispensable good.[83] It is also the heart of the bioregional vision. Its design begins with teaching children the stories and facts about where they live and with whom they cohabit, instilling in them the knowledge that they are participants within a region and the planet. Teaching immersion in the rhythms and amidst the beings of place allows for a sense of membership in the natural world to take root early in the human psyche. Indigenous people understood the value of teaching intimacy with nature. Here are the words of Chief Luther Standing Bear:

Kinship with all creatures of the earth, sky and water was a real and active principle. For the animal and bird world there existed a brotherly feeling that kept the Lakota safe among them and so close did the Lakotas come to their feathered and furred friends that in true brotherhood they spoke a common tongue.

The old Lakota was wise. He knew that man's heart away from nature becomes hard; he knew that lack of respect for growing, living things soon led to lack of respect for humans too. So he kept his youth close to its softening influence.[84]

The virtues of kindness, compassion, care, and restraint are cultivated in nearness with the natural world and become organically acquired as extensions of human nature. The Karuk tribe of the Pacific Northwest expressed the same understanding as the Lakota:

Close, lifelong observation of nature and the landscape is much admired in the Karuk culture. Indian children have early on been placed in a life trajectory which establishes, at the deepest levels of the mind, habits of quiet observation and a sense that they are not necessarily in control of the world around them. When people's actions are influenced by keen observation of nature, they are much less likely to attempt to dominate or desire to change natural processes.[85]

Humans, of course, do and must change the world: but when attentiveness to the more-than-human world is taught and cultivated, when children are initiated into *belonging* as "the pivot of life,"[86] people (will) change the world in respectful, appropriate, and even beautiful ways.

Every human being brings a gift when they come into the world. The paramount objective of education is to offer the conditions and learning environment that can bring each person in contact with that gift, that calling, that transcendent staircase into their innermost being. What those conditions include will vary from bioregion to bioregion, but universal among them is clean, hearty, and healthy nutrition. Undernourishment, malnourishment, and hidden hunger trail cognitive and emotional deficiencies, which are endemic in the current food system, affecting not only the globe's poorest people but the world's overfed as well.[87] This execrable situation of global food inequity and rampant bad food is a most fundamental corrective target of bioregional societies. Feeding children and all people well and teaching them holistic care for their bodies are imperatives of an enlightened ecological civilization.

Education will reach beyond formal schooling of elementary and secondary and professional and doctoral degrees, with an available "capstone" in what Nash calls the University of the Wilderness.[88] Skills and arts for journeying into the wild, as well as the practices of sailing, snorkeling, and scuba diving for exploring the seas, will be supported by expert mentors and outward-bound teams for those wanting to avail themselves of such education. This kind of wild learning is patterned on the quest for self-transformation and the pursuit of higher truths through the elemental fire of challenging experience. It provides the soul nourishment of immersing one's whole being in the primordial beauty of the biosphere. Environmental philosopher Alan Drengson calls adventuring in the wilderness the "Wild Way" and describes it as a "learning and practice system" that combines "unstructured wandering" with mindfulness and whole-body disciplines of different cultures, such as Eastern martial arts, Indian yoga, and the shamanic traditions of native people.[89]

Bioregional identities will differ from place to place yet lifestyles in each bioregion will exhibit a "creative bricolage," meaning, as Taylor explains, "an amalgamation of many bits and pieces of diverse cultural systems."[90] Through global connectivity, humanity has created an invaluable diaspora of cultural motifs and practices—all of which may occur in any bioregion. For example, the Appalachian bioregion of North America will have its native traditions of food, dance, music, recreation, and the arts, but in it one might also encounter Japanese gardens, Inca-style terracing for food cultivation, Himalayan fusion eateries, African drumming practitioners, or Taoist meditators living remotely. Cultural shards will coalesce uniquely and fluidly in different bioregions, even as the primary unit of cultural diversity may become the *individual*. "The best life possible," argued social historian Lewis Mumford, "is one that calls for an ever greater degree of self-direction, self-expression, and self-realization. In this sense, personality, once the exclusive attribute of kings, belongs on democratic theory to every man. Life itself in its fullness and wholeness cannot be delegated."[91]

Nothing promises greater potential of human freedom than the availability of conditions for each person to self-create themselves by means of tools, ideas, practices, and traditions that humanity has developed and explored in a great variety of cultures. Such a universal vision of cultural exchanges and collages is captured as "cosmopolitanism" in Western political philosophy. Cosmopolitanism is, itself, a great contribution of Western culture to humanity, "signifying an attitude of enlightened morality that does not place 'love of country'"[92] ahead of love for all Earth's people and nonhumans. Cosmopolitanism "signifies hybridity, fluidity, and recognizing . . . that the complex aspirations [of human beings] cannot be circumscribed by national fantasies and primordial communities."[93]

Every bioregional settlement will of course have its own cultural "feel." Far from being provincial and xenophobic, however, the bioregional way of life sides with the cosmopolitan aspiration of open communication and solidarity between peoples—a transcultural conversation of humanity—based on mutual trust and respect.[94] There is little prospect for world peace without the universal institutions and norms to make this conversation a down-to-earth reality. Therefore, a global ecological civilization—a federation of bioregions, if you will—must transcend every stripe of nationalistic delusion, religious fundamentalism, and cultural or ethnic zealotry, all of which "restrain the human mind within the smallest possible compass, making it the unresisting tool of superstition, enslaving it beneath traditional rules, depriving it of all grandeur and historical energies."[95] Preserving the unique contributions of every culture—while forfeiting the oft-attendant downside of xenophobia or worse—can be achieved when "the individual" becomes the fundamental unit of cultural diversity, as self-creative bricolage from humanity's sundry heritages. Far from contradicting cosmopolitanism, bioregional inhabitation shapes its highest expression by grounding it in love of place. Indeed, bioregionalism expands the cosmopolitan vision beyond its conventional anthropocentric straightjacket, to include the wellness of all earthlings, their homelands, and conditions they require to thrive.[96] Bioregions will thus be doubly nestled within Earth's wild expanse and within a global civilization characterized by the openness demanded by human beings' enormous breadth of potential and for realizing peace between all peoples.

Cosmopolitanism implies a world with porous human borders. To sustain freedom of human movement, alongside a dynamic equilibrium of bioregional communities, means that each bioregion will cultivate, sustain, and cherish its own wealth—wealth flowing from the beauty and gifts of its own "naturecultures."[97] Without entrenched structural inequalities between wealthier and poorer bioregions, the movement of people will follow personal predilections, desire for learning, and fallings-in-love. Even as human migrations will be relatively unrestricted they will not be torrential or restless, for human beings will be raised in deep familiarity with the places of their birth. Intimacy with place precludes the hypermobility of today's modern, rootless people. With every bioregion cultivating a high quality of life within nature's abundance, unbalanced migrations of people will not arise. In that sense, cosmopolitan bioregionalism aligns with the preference of "staying put" that bioregional thinkers have long emphasized: Gary Snyder entreated his fellow men and women to dress up and stay home, while consummate bioregionalist Henry Thoreau lived his passion of getting closer and closer to where he was.

The bioregional ideal is about becoming native to place, intimate with its ecologies, beings, history, and other unique features. As two of the original theorists of bioregional life, Peter Berg and Raymond Dasmann, stated, bioregionalism is "becoming fully alive in and with place."[98] Wendell Berry writes that it is "local love, local loyalty, and local knowledge that make people truly native to their places and therefore good caretakers of their places."[99] There will be no such thing as a declining ecological baseline or ecological amnesia within bioregional communities, since human life will be attuned with the prospering of all native and established species, citizen sciences will proliferate, and bioregional cultures, emulating indigenous ways, will create celebratory traditions for honoring and committing to memory all its denizens. Like the Aboriginal people of Australia, the citizens of bioregional communities "will sing up the country," celebrating the participatory nature of all-species world-making.[100] Bioregionalism is thus grounded in an ethos of kinship with the living world that is both ancient and modern. Like the mythical serpent Ouroboros turning around to bite his own tail, the modern sciences of taxonomy, ecology, and evolution return (while adding a scientific twist) to the indigenous knowledge of interdependency and relatedness of all earthlings.

Bioregional polities will splinter or cede out of nation-states, but their geopolitical identity is entirely another thing. Bioregions are designed to fit within landscapes, within their living and productive features, not to dominate as the current human-supremacist status quo imposes. Bioregional societies will enact decentralized politics, meaning they will govern themselves at a human scale. Big government and unnecessary bureaucracy, as well as the office of the "professional politician," will be supplanted by a political process that is fully transparent. Democratic governance will realize its inherent validity claim—turned into a sham everywhere by politicians and established political practices—of serving the denizens and the long-term health of place. Bioregional societies will guard themselves against political and economic corruption, which in any case will be all but structurally precluded from the fact that bioregionalism's very emergence implies superseding the idea of nature as a suite of "resources." Without that keystone anthropocentric creed—the ideological prerequisite for nature to be convertible into currency and stuff—neither a consumerist culture nor gross economic inequality can find a foothold. In the absence of disempowered groups of humans to do the slaughterhouse, janitorial, assembly line, and other menial work, the entire structure of any remaining psychologically harmful, repetitive, and/or mind-numbing work will have to be rethought: it will certainly not involve invisible and fungible people with shorter than average lifespans.

We can already discern incipient tendencies toward bioregional re-inhabitation, for example, in intentional communities, ecovillages, and transition towns. If these trends are a sign, bioregionalism will likely emerge more through revolutionary implosion than direct confrontation, as people who become revolted or disenchanted with dominant culture, fed up with inane nationalisms, corporate corruption, and anthropocentric abuses seek to regroup into new social identities around the worthy gifts of cultures and in solidarity with Earth's diverse lifeforms. As political thinker Luke Philip Plotica puts it (describing Thoreau's politics but valid more generally), "rather than evasion of politics or social responsibility . . . withdrawal is a deliberate response to the complicity of ordinary life in conditions of moral and political degeneration."[101] Instead of fighting nation-state and corporate power for "privileges" invariably framed in power's terms, a bioregional movement can show its backside to power by a performative politics of exodus from conventional society. Human beings everywhere can build another way of life by engaging "the exercise," in Gary Snyder's words, "of ignoring the presence of the national state,"[102] while simultaneously opting out of consumer culture and a fossil-fuel-based economy.

To avoid recycling past historical mistakes, bioregional communities must grapple with reinventing human life at deep levels, including, for example, designing novel non-anthropocentric geographical maps; exorcizing the late-modern crimes against humanity that critical theorists have railed against as the enslavement to speed, the annihilation of time, and the permanent electrocution of the human nervous system by the infinite velocity of cyberspace and the ubiquity of screens;[103] and, not least, deracinating human-supremacist constructs out of language. Such linguistic constructs include conventionalized slurs, put-downs, objectifications, and dualisms, such as "savages" and "natural resources" and "higher" versus "lower" animals, as well as commonplace expressions like "pests" and "weeds."

The language we bring to the world is of paramount significance, for language is a primal dimension. While it has been exquisitely shaped by nature (and natural selection), language has contouring power over reality. "Language speaks," as has been famously stated, to convey its sovereignty over the speaking subject.[104] If humans do not wish to be the unsuspecting instruments of language, guided unconsciously by its insinuated dictates, then what language is saying when "it speaks" must be closely heeded. Fortunately, we have the capacity to listen closely to language by means of what one Western philosopher called "the piety of thinking"[105] and Eastern wisdom traditions describe as "the witness mind." By shining critical awareness on concepts and turns of phrase, their gravitational pull can be discerned: if their gravitational pull lands us in a world we do not wish to inhabit, we have the power to reject the language that takes us to, or tugs us toward, that world.

Naming itself has conjuring force, shaping perception and guiding action. Therefore, citizens of an ecological civilization will attend to the importance of naming beings with care. Beings and places must not be saddled with names that speak the language of utility, disservice, objectification, or human ownership, such as when farm animals are called "livestock," wolves and coyotes are called "vermin," or fish are called "fisheries." All naming is, of course, embedded in the human sphere, because naming is what humans do. When a creature or place is named with the intention to name it for what it is, no harm will follow; calling a coyote a "coyote" is such a naming. When wolves or coyotes are called "vermin," however, evil becomes unleashed on the world. In the heyday of wolves being called vermin in North America, people did monstrous things to wolves like setting them on fire or tying their mouths shut and letting them go. The human-supremacist worldview devises such language (like "vermin"), but language itself has the power to maneuver perception into the modality of *seeing as*.[106] Awful things can be done to beings who are seen as vermin. In recent times, we have also witnessed extreme cruelty, extermination campaigns, and herbicide warfare directed against animals and plants labeled "invasive."

The power of language to cast a dark net over the world, its power to bewitch and lead morally astray, cannot be overestimated. When fish are called "fish," no harm follows. When fish are called "fisheries," almost anything goes. Fisheries is a thoroughly instrumental and human-possessive concept, but the violation potential of the concept is magnified by the fact that the concept steers people to see fish as fisheries. Everything else gets wiped out of view: the fish as beings-in-themselves and forms of life, the deprivation of other marine creatures of food, the sheer wealth of ocean life, the importance of the sheer wealth of ocean life to cocreate the whole world. A universe of life fades and a facade of fisheries reigns. Beyond withering the human imagination, the designation of fisheries sponsors a violence of monstrous proportions, while simultaneously hiding the violence behind the conjured veneer of human ownership. Like "vermin," "fisheries" places blinkers on the human mind, which subsequently fails to see either the beauty of the world or the violence of the human hand. Seeing the natural world as "natural resources" acts likewise. How ironic! Language has always been the favorite Rubicon of Difference between humans and all else, yet that (purportedly) unique feature itself has the power to trap people into a way of seeing that can lead them by the nose.

"A picture held us captive," wrote philosopher Ludwig Wittgenstein, "and we could not get outside it for it lay in our language and language seemed to repeat it to us inexorably."[107]

Language is a major shaping force of the cognitive and emotional frames through which humans perceive, comprehend, and feel reality. Indeed, language often becomes the all-powerful shaping force, when its framings are taken on board wholesale, and parroted and lived uncritically—in other words, when conceptual renderings are projected onto reality as if reality itself had offered them up and molded them into language. When a river becomes conceived as "freshwater," for example, a universe of living beings, and their cocreation of the river as a living world, are erased and delivered into violence and oblivion, and the dwelling places of rivers get conceptually twisted into "natural resources" as though "natural resources" are actually out there.

Bioregional societies will take care to purge language of the human-supremacist constructs that, in their endless repetition, hold present-day humanity captive to that worldview. Or perhaps those linguistic constructs will simply fall into disuse as their ugly ring becomes obvious, just as virtually no one today is inclined to call indigenous people "savages."

The reason that vermin, fisheries, weeds, livestock, and even invasive species can sponsor individual and/or conventional crimes is because—as pure referents to disservice, human ownership, killability, or objectivized utility—they smother being-in-itself, and give rise in the human mind and collective to oblivion of being-in-itself. Yet the being-in-itself remains—it is real—it does not go away because it can be shot from a helicopter, poisoned with 1080, obliterated with trawlers, carpeted in herbicides, or reproductively subjugated for a (short and brutal) lifetime to mass-producing dairy. The being-in-itself has not gone away no matter what is done to it, but oblivion and the darkness of oblivion reign. Human-supremacist language works to conceal violence, suffering, destruction, and cruelty—so that these inflictions transpire unperceived. By the same token, what transpires for the human (perpetrators and their complicit societies), also without being perceived, is what novelist J. M. Coetzee[108] captured in a well-chosen thorn of a word: *dis-grace*.

The language of oblivion—the concepts that conceal the being-in-itself of all life—will be purged and fall into disuse in an ecological civilization. This will support the grace to flow from being present, with clarity and openness, before the being of all life. For the being-in-itself of all life is the greatness of life as such.

Within a federation of bioregions, the biosphere's large-scale, reconnected wild expanses will exercise their own forms of self-governance as much as the bioregional communities ensconced within them. Wild animals will repatriate and repopulate the world, migrations will flourish and renew their rhythms, forests will grow old and resplendent, rivers carve their life-filled paths freely, and processes of speciation for all life's taxa along with conditions of ecological complexity will become reinstated. The restoration of forests and grasslands will play a vital part in absorbing the excess carbon dioxide unleashed after the Industrial Revolution.[109] Along with the need to shift to distributed and renewable energy sources, continental- and oceanic-scale restoration and rewilding will help to bring the menace of rapid climate change under control.

It is the big rhythms of Earth's forests, grasslands, savannahs, mountain and seamount ecologies, wetlands, deserts, freshwater bodies, coral reefs, continental shelves, and high seas, with all their exuberant biodiversity, that form the milieu of integral human inhabitation; for "the homeland of all humans is the whole planet rather than some piece of it."[110] Such a newfound inhabitation by humanity presupposes a new consciousness—one that blends evidence-based reasoning about the world, the indigenous wisdom of balanced living, compassion for all living beings, and safeguarding the memory of awe for existence itself. This may sound impossibly idealistic or utopian. Even so, the bioregional vision is grounded in reality—the reality of natural laws, relationships, countless beings, embeddedness, and interdependence, along with all beauties and riches spun therein.

Utopian is a way of life that remains stubbornly disconnected from and deaf to the full gamut of demands that being entangled within a living cosmos places on us. Not facing the full gamut of demands is the primary ingredient of the rocket fuel powering the ships headed for "utopia." This is why the prevalent technological euphoria, and its sidekick, the technological fix, comprise the actual utopian, albeit avidly pursued, program of our time. Desalination, nuclear power, artificial intelligence, nanotechnology, synthetic biology, geoengineering, the genetic engineering of whatever, space travel, the singularity, and more are due to come to the rescue—or even transport humanity to a brand-new brave new world. Even as an expression of hope all this technological rapture is superficial, because its execution is founded on Earth's colonization, which technological prophets remain silent about, refuse to acknowledge, or appear completely unaware of.

The technological program of salvation saves nothing and no one except (possibly) an upgraded version of "the human enterprise." The flight into technological redemption is utopian not because its sundry promises are physically unrealizable (they may or may not be), but because they are not grounded in the fullness of the real. The technological approach reduces the gamut of demands that being entangled in a living cosmos places on us to a set of technical and economic questions. Thus, for example, entrepreneur Elon Musk's dream to make humanity a "multiplanetary species" (with the colonization of Mars for starters)[111] is disconnected from reality not because its realization faces intractable technical and economic hurdles, but because this dream is presented as one that *only* faces technical and economic hurdles. Rome is on fire, but the Wagnerian Ring Cycle of humanity's technological valor resounds, in endless variations, over the crackle and the smoke.

Epilogue: Toward an Ecological Civilization

The ecology movement, to the extent that its central worry is the rapid
extinction of ecological diversity, is essentially a resistance movement against
the imperialism of human monoculture, roughly analogous to the earlier
resistance movements against particular totalitarian regimes. JOHN RODMAN

We live in a time of ecological awakening, yet often find
the urgencies it articulates to be misstated. Warnings such
as "we now face a global crisis in land use and agriculture
that could undermine the health, security, and sustainability of our civilization"[1] recur in the environmental
literature. The plea "to save civilization" from the global
ecological crisis that civilization itself has unleashed is a
common theme. Far from needing saving, however, civilization must be dismantled so that Earth's endangered life
can be preserved and restored and the opportunity to create a more enlightened civilization can emerge. Civilization, as we know it, rests squarely on the domination of
nature: as I have argued in this work, domination is not
a contingent aspect, correctable defect, or unfortunate
side effect of civilization, but constitutional to its very
character.

Even if a global civilization were to succeed, on its own
techno-managerial terms, in resolving the challenges confronting humanity, it could not avert the existential and
ethical consequences of its subsequent reign. Its very success entails those consequences: the extinction of species,
subspecies, and populations of plants and animals; the

destruction and simplification of ecologies; the loss of diverse behaviors, cultures, and forms of awareness among nonhumans; the reduction of biodisperity, or the uniqueness of varied places on Earth; the elimination of the requisite genetic diversity for life's full evolutionary dynamism; the dwarfing of wildlife populations by human and domestic animal biomass; and the proliferation of land and sea factory farms "to feed the world." A victorious civilization will shrink a biodiverse world and replace it with a humanized one, checkered with all manner of development and industrial infrastructures, displaying homogenized ecologies with the same cadre of animal and plant species recurring globally, and depending on the enslavement and exploitation of living beings for its very continuity and for the production of what it defines as "wealth." Civilization is in process of producing such an occupied world, with nonhumans disciplined into nature reserves or subjugated with normalized cruelty in zoos, research labs, and CAFOs.

The focus of much environmental analysis on the frightful potential of civilization's collapse is amiss, for it implies that what is wrong with our ecological predicament is that it is endangering civilization. The collapse thesis tends to rally people around the distorted mandate of "saving civilization," instead of redirecting our energies toward protecting a living planet and all its beings from civilizational plunder. Civilization does not need to be saved, but reinvented.

A total of roughly fifteen thousand nuclear weapons exist in the world.[2] That fact alone (a sliver of global military capacity) is a sufficient pointer that something has gone seriously awry in the very imaginary of history's course. Moreover, military spending is on the rise around the world, and most especially the United States. "When the biggest crises facing the planet require education, training, health care and investment in sustainable energy and agriculture," points out author Raj Patel, "governments are piling up record sums into guns, not [food]."[3] We should reasonably dread where such priorities are heading the world. What is fundamentally criminal about the kind of civilization that has hijacked human history, and continues to hold humanity hostage, is that it is founded on taking over whatever aspects of the nonhuman world are needed for conversion into wealth and commodities, while displacing and killing whatever aspects of the nonhuman world stand in the way of appropriation. By logical and pragmatic extension, civilization is also founded on obliterating whatever aspects of human worlds (cultures or individuals) obstruct, for whatever reason, access to resources. (For example, the headline of a 2017 *Guardian* report reads, "Environmental Defenders Being Killed Globally in Record

Numbers, New Research Reveals."[4]) The bedrock of nature colonialism on which civilization stands has built perpetual violence into its very edifice. A horrific culmination of this historical juggernaut lies in the inane and dissociated fabrication of pan-catastrophic weapons that imperil all of humanity and the entire biosphere.

We should neither fear nor hope that nature (for example, catastrophic climate change) will bring civilization down. We should strive to bring it down ourselves. It is not worthy of the highest aspirations of the human: should civilization's nature-parasitizing project be successfully sustained, humanity will have arrested its moral evolution; and should civilization's nature-parasitizing project fail, humanity will fall into a pit of conflict, violence, and suffering that is unbearable even to contemplate.

An immediate turn in the direction of a global ecological civilization is therefore the only real option. It is impossible to foresee what such a civilization will look like; that will be a work in progress for future generations to shape. What we know at this historical juncture is that moving in that direction entails opposing "the trends of more," rather than striving to accommodate and salvage them by means of piecemeal interventions, fixes, technological gigantism, or plain muddling through dire events and chronic problems. The burning question is not whether Earth can support over ten billion people with an industrial diet, technologies, mobility, and consumer comforts. The point is that succeeding to provision billions of people with such a lifestyle means entrenching the domination of nature and the subjugation of nonhumans, while simultaneously vastly diminishing Earth's biological wealth.

When human beings become free to think and act without identifying with the supremacist worldview, they will break a chain that has linked countless generations.[5] Then the baton of human supremacy will be incinerated, effecting "a historical break from the continuum of domination" and witnessing the emergence "a new type of man."[6] Why is it, we might reasonably wonder, that through numerous social revolutions that have sought to establish justice for all humanity, "the continuity of domination has been sustained"?[7] Is it because the human will to dominate other humans is hardwired into our genome thereby always reasserting itself? How very unlikely. In line with what I have highlighted throughout this work, the will to dominate reasserts itself because at the core of all human-human domination lies the worldview of nonhuman nature as "resources" for wealth and power. If the human-supremacist worldview is not toppled, social injustice and inequality will continually rehearse themselves in one form or another. As environmental analyst David Johns put it, "human efforts to control Nature [require] forms of social organization and technology that at heart involve the control of some humans by others."[8]

No social movement agitating for "liberty, equality, and fraternity" can succeed as long as the constitution of the biosphere as humanity's colony reigns. Social relations never transpire in a vacuum, despite the fancy that anthropocentrism has long cultivated that the natural world is just the backdrop for the all-important show of human affairs. It is within the context of the dominant relationship between humanity and Earth that social relations (relations between people) have become constituted as material, normative, and historical realities. As long as the living world is construed as a container of resources, social relations will more or less express the corollaries of this belief: there will be competition, exploitation, corruption, struggle for access and control, posturing, and war over all manner of resources. As Jack Turner states the case, "when 'the world' shrinks into a rationalized grid stuffed with resources, greed goes pandemic."[9]

Inequalities and distortions in human relations are inexorably tied to the mindset of resourcism. They are inflamed by the enacted regard of the natural world as a domain to be used for human profit or advancement. To state the same idea more bluntly, borrowing again from Turner, a dominant culture that "treats Mother Nature like a whorehouse"[10] will not exactly cultivate human virtue or compassionate character among human beings. The fountainhead of the disparity between the haves and the have-nots, between the powerful and the powerless, lies in the conception and treatment of Earth's living beings and nonliving things as sources for aggrandizing wealth and power. Social justice is thus not achievable as long as the natural world continues to be stripped of its intrinsic standing and reconfigured as a collection of dead and killable stuff. By virtue of the constructed entities they are, resources encourage and produce the acquisitive mindset that undergirds human conflict, corruption, and injustice. "What impoverishes nature also impoverishes man," wrote John Rodman, "in terms of the type of human being that is produced in the course of a life devoted to transforming nature into commodity."[11]

On the other hand, in the unlikely event that people achieve greater equality—while sustaining the constitution of nonhuman nature as made for humans—then a more equitable "distribution of resources" will come to pass at the expense of Earth's biological plenum and of nonhuman freedom and flourishing. Were resources more equitably distributed, what kind of justice would prevail if the bounty of ecologies at landscape and seascape levels was destroyed, uncountable species were extinguished or corralled into zoos and reserves, and livestock and fish were reduced to factory-farmed objects for turning into "protein"? The hypothetical attainment of an anthropocentric social-democratic (or socialist-type) system, guaranteeing justice only for humans, would render the very concept of justice specious and shallow.

In agreement with philosopher Gary Steiner, the human struggles for justice and recognition are "urgent and entirely real." He goes on to add: "The ideal of cosmic justice," wherein we see ourselves as inhabitants of Earth and the cosmos, "seeks not to devalue those struggles but rather to place them in a larger world context that we have tended to repress from the beginning of civilization to the present."[12] Because of this repression—of turning a blind eye to the apartheid between human and nonhuman—the history of social inequality and injustice has repeated itself for millennia. Inequity between people has been an essential component of civilization parasitizing on the more-than-human world since the days of Gilgamesh, the king of Sumer, who had the Cedar Forest cut down because, in his own words, he wanted to be rich and famous. (Rich and famous he became and still died a broken man.) Humanity will exit the cycle of repetition only if we revamp civilization so radically that the way into the future cannot be "construed as a sequence belonging to the process of history."[13] It is the most heroic task human beings have ever been called to—and a task we are capable of rising to, for there are so many reasons to wager that under the accreted armor of human supremacy lies an ocean of tenderness and love for the living world.

Notes

INTRODUCTION

1. Leopold (1949) 1968, 149.
2. Invisible Committee 2009, 14.

CHAPTER ONE

1. deBuys 2015.
2. Smil 2011, 618.
3. Quoted in Roberts 2007, 119–20. "Grampus" is likely a word for killer whale.
4. Roberts 2007, 119.
5. The word "herring" is thought to originate from the German word for army. Roberts 2007, 119.
6. Waldman 2010; Miller 2005.
7. Quoted in Sale 1990, 310.
8. Darwin (1859) 1964, 490.
9. Ceballos et al. 2010; Mora et al. 2011. These estimates do not count microorganisms, which if included would bring the total closer to one hundred million.
10. Wilson 2015.
11. Ceballos et al. 2010.
12. Annie B. White, "Gray Wolf Conservation: A History of Wild Wolves in the United States," http://www.graywolfconservation.com/Wild_Wolves/history.htm.
13. Dinerstein 2016, 16–24.
14. Dirzo and Raven 2003, 144. Wolves around the world numbered in the millions, but persecution and destruction of habitat has reduced their numbers to one-tenth of what they were. See also Ceballos et al. 2015, 91.

15. Dirzo and Raven 2003, 142–43.
16. Ehrlich and Pringle 2008, 11580.
17. Wilson (1999) 2010.
18. Barnosky et al. 2011.
19. Ceballos et al. 2010.
20. Wilson 2002; Dirzo and Raven 2003, 164.
21. Barnosky et al. 2011, 51; Kolbert 2014.
22. Whitty 2007, 38.
23. Darwin (1859) 1964, 126.
24. Myers and Knoll 2001, 5389.
25. Rolston 1985, 723.
26. Ceballos et al. 2010.
27. Wilson (1999) 2010, 243.
28. Pimm et al. 2014, 2. See also Dirzo and Raven 2003, 161; Lees and Pimm 2015.
29. Dirzo and Raven 2003, 154.
30. Lees and Pimm 2015, 179.
31. Quoted in van Dooren 2014, 7.
32. Ceballos et al. 2015, 22; Ehrlich and Pringle 2008.
33. Ceballos et al. 2010.
34. Maxwell et al. 2016.
35. Ceballos et al. 2015, 85.
36. Dirzo et al. 2014; McCauley et al. 2015.
37. Rebecca Morelle, "World Wildlife 'Falls by 58% in 40 Years,'" BBC News, 2016, http://www.bbc.com/news/science-environment-37775622. See also Spash 2015.
38. See Crist 2013a.
39. Ceballos et al. 2015, 148.
40. Duncan 2013, 13.
41. Pimm 2013, 56.
42. Lees and Pimm 2015, 177; Ceballos et al. 2015, 123.
43. Ripple et al. 2015, 5.
44. Wilson 2006, 80. Nicole Battaglia, "The Status of Haiti's Fisheries: When Numbers Don't Match Reality," Environmental Performance Index, Case Study, 2013, http://archive.epi.yale.edu/case-study/status-haitis-fisheries-when-numbers-dont-match-reality.
45. Manning 2004.
46. Drop a "thank-you" note for that to Monsanto, recently in the process of merging with Bayer, the company that developed the neonicotinoid pesticides implicated in colony collapse disorder.
47. Quoted in Cornwall 2014.
48. Ceballos et al. 2015, 69.
49. Polidoro et al. 2010.
50. Polidoro et al. 2010.

51. Sanderson et al. 2002, 893.
52. Mills 2010.
53. Laurance 2015; Laurance et al. 2015.
54. Whitty 2007.
55. According to the International Union for Conservation of Nature, "a protected area is a clearly defined geographical space, recognized, dedicated and managed, through legal or other effective means, to achieve the long-term conservation of nature with associated ecosystem services and cultural values." See Wuerthner, Crist, and Butler 2015.
56. Pimm et al. 2014, 5.
57. Pimm et al. 2014, 6.
58. On "paper parks," see Terborgh 1999. On the tragedy of Southeast Asia's poaching crisis, see deBuys 2015; Stokstad 2014.
59. Maxwell et al. 2016; Ripple et al. 2016.
60. Danson 2011, 130–35.
61. Ripple et al. 2016; Estrada et al. 2017.
62. Lee et al. 2014; Petersen 2015; Shukman 2016.
63. Ceballos et al. 2015, 124.
64. Jones 2010; Ripple et al. 2016.
65. Ripple et al. 2016.
66. Ripple et al. 2016.
67. Ripple et al. 2016, 9.
68. See Chris Shepherd, "Trade Wiping Out Indonesia's Bird Species," TRAFFIC: The Wildlife Trade Monitoring Network, 2016, http://www.traffic.org/home/2016/5/25/trade-wiping-out-indonesias-bird-species.html.
69. Williams et al. 2015, 9.
70. Wilson 2006, 36, 47.
71. Ceballos et al. 2015, 30, 43–44.
72. Ceballos et al. 2015, 30, 43–44.
73. Dirzo and Raven 2003, 160.
74. Duncan 2013, 10.
75. Diaz and Rosenberg 2008; Hance 2008.
76. Ceballos et al. 2015, 73.
77. Wilson 2006, 78.
78. Lenzen et al. 2012, 110.
79. Madhuri et al. 2014, 17.
80. Morelle 2016.
81. Mills 2010, 6.
82. Kottasova 2016.
83. do Sul and Costa 2014, 360.
84. See van Dooren 2014, chap. 1.
85. "It's a Plastic World," https://vimeo.com/100694882.
86. Casey 2010, 14.
87. do Sul and Costa 2014.

88. Urban 2015; Jaworowski 2015.
89. McKibben 2011, 42.
90. Pounds et al. 2006, 165.
91. McKibben 2011, 25.
92. Morello 2010; Danson 2011, 64–65.
93. McKibben 2011, 10.
94. Noss 2001.
95. Root and Schneider 2006, 708.
96. Stephens 2010; Huey, Losos, and Moritz 2010, 832.
97. Maxwell et al. 2016.
98. Barnosky et al. 2011, 56.
99. Weisman 2007, chapter 19.
100. Whitty 2007.
101. Monastersky 2014.
102. Ceballos et al. 2015, 55.
103. Sodhi et al. 2004; Ripple et al. 2015, 5.
104. Ceballos et al. 2015, 57.
105. Ceballos et al. 2015, 109; Ripple et al. 2016.
106. Corlett 2016; Stephen Buchmann, "Our Vanishing Flowers," *New York Times*, editorial, 2015, https://www.nytimes.com/2015/10/17/opinion/our -vanishing-flowers.html.
107. Ceballos et al. 2015, 61.
108. Quoted in Lovgren 2006.
109. Dudgeon 2010.
110. Quoted in Lovgren 2006.
111. Böhm 2013, 378, 381.
112. *Nature* 467 (September 9, 2010): 136 (news briefing).
113. Ripple et al. 2016, 2.
114. Whitty 2007, 45.
115. Myers and Knoll 2001.
116. Jamieson et al. 2017.
117. For example, the US Endangered Species Act has saved over two hundred species from extinction (Ketcham 2017a). See also Goodall 1999; Hawken 2007; Seddon et al. 2014; Hannibal 2016.
118. See Dictionary.com, s.v., "epic," http://www.dictionary.com/browse/epic.
119. Hawken 2002, 11.
120. Ehrlich and Holdren 1971.
121. Ehrlich and Harte 2015.
122. Barnosky et al. 2012, 56.
123. Ryerson 2015.
124. Foley et al. 2011.
125. Maxwell et al. 2016.
126. Sanderson et al. 2002, 895.
127. Foley et al. 2005.

128. UNEP 2012.
129. Hance 2008; Diaz and Rosenberg 2008.
130. Tilman et al. 2001.
131. Hoekstra et al. 2005.
132. Gibbs et al. 2010.
133. Butler 2008.
134. Polidoro et al. 2010.
135. Dudgeon et al. 2006; Vörösmarty et al. 2010.
136. Estrada et al. 2017.
137. Jackson 2008.
138. Danson 2011.
139. Hoekstra et al. 2005.
140. Myers and Worm 2005; Roberts 2007; Erle 2009; Smith et al. 2010; Jackson et al. 2011; Danson 2011; Kurlansky 2011; Gjerde et al. 2013; Stone 2013.
141. Ripple et al. 2015.
142. Wilcove 2008a.
143. Estes et al. 2011; Ripple et al. 2014; Bauer et al. 2015.
144. Quoted in Biello 2009a.
145. Manchovina et al. 2015.
146. Halweil and Nierenberg 2008, 61.
147. Bittman 2008.
148. Weis 2013, 80.
149. Growing crops to feed animals to feed people is tremendously inefficient by comparison to eating plants or grains directly.
150. See, for example, Miller 2017.
151. Ripple et al. 2015, 8. See also Manchovina et al. 2015.
152. Ripple et al. 2015, 3.
153. McKibben 2010.
154. Lenzen et al. 2012.
155. Lenzen et al. 2012, 111.
156. Ripple et al. 2015, 4–5.
157. Duncan 2013.
158. Ellis 2012.
159. Weisman 2013, 95.
160. Plumwood 2008.
161. Lanzen et al. 2012, 111.
162. Blackwelder 2015.
163. See "View from the Bridge," *Economist*, 2013, http://www.economist.com/news/finance-and-economics/21569729-what-big-american-port-says-about-shifting-trade-patterns-view-bridge.
164. "View from the Bridge," *Economist*, 2013.
165. "View from the Bridge," *Economist*, 2013.
166. Hansen and Gale 2014.

167. Vijay et al. 2016.
168. See https://www.rainforest-rescue.org/topics/palm-oil.
169. Ceballos et al. 2015, 102.
170. Ceballos et al. 2015, 110–12.
171. Wilson 2006, 75.
172. Vijay et al. 2016.
173. Schor 2010.
174. Quoted in Weisman 2013, 116.
175. Wilson 2006, 54.
176. Ehrlich and Pringle 2008, 11580.
177. Suckling 2014.
178. Ceballos et al. 2010; Ehrlich and Ehrlich 2013; Ellis 2012.
179. Wilson 2015.
180. Ehrlich and Pringle 2008, 11580.
181. Rose 2013, 92.

CHAPTER TWO

1. For analyses of anthropocentrism, see Thomas 1983; Metzner 1993; Sessions 1995; Chew 2001; Eckersley 2005; Steiner 2010; Boddice 2011; Schroll and Walker 2011; Curry 2012; Calarco 2014; Kidner 2014; and Jensen 2016.
2. Curry 2012, 55.
3. Marcuse 1969, 11.
4. Thomas 1983, 24; Noske 1997.
5. Marcuse 1969, 11.
6. See Washington 2018.
7. Evernden (1985) 1999, 48.
8. Habermas 1984, 61.
9. Huxley (1889) 2012.
10. Darwin (1871) 1981, 405.
11. Duncan 2013, 4.
12. Darwin (1859) 1964, 128.
13. See Forbes (1979) 2008.
14. Louv 2008.
15. See Redman 1999; Chew 2001; Wright 2005; Ponting (1991) 2007.
16. Sax 2011, 31.
17. Haven 2006, 18.
18. Krall 2007, 137.
19. Krall 2015.
20. Quoted in Jonas 1974, 122.
21. Rodman 1980.
22. Steiner 2010; Cavalieri 2006.

23. Patel 2011, 101.
24. Rodman 1980, 54. See also Nimmo 2011.
25. Renaissance humanist Leon Battista Alberti, quoted in Sale 1990, 38–39.
26. Humanist Marsilio Ficino, quoted in Sale 1990, 39.
27. Calarco 2012, 58.
28. Agamben 2004.
29. Crist 2013a.
30. Derrida 2008, chap. 1.
31. Thomas 1983, 31.
32. Oliver Goldsmith, quoted in Thomas 1983, 35.
33. Lovejoy (1936) 1976.
34. Lovejoy (1936) 1976; Thomas 1983; Sax 2011.
35. Buber (1970) 1996.
36. Thomas 1983, 46–47.
37. Thomas 1983, 47.
38. Calarco 2014, 418.
39. Campbell 1964, 240.
40. Debord 1995, 29.
41. Plumwood 2008.
42. For perspectives on indigenous worldviews in relation to the natural world, see Mander 1992; Mander and Tauli-Corpuz 2006; Descola 2009; Rose 2013, 2014.
43. Three hundred fifty parts per million is the scientifically estimated maximum level of carbon dioxide in the atmosphere that avoids dangerous climate change. The atmospheric composition of greenhouse gases today exceeds four hundred parts per million and continues to rise. See McKibben 2012, 2016.
44. Horkheimer and Adorno 1972, 9.
45. Waldman 2010. See also Jackson et al. 2011.
46. Safina 2011, 18.
47. Rodman 1987, 94.
48. Heidegger (1947) 2013, 291.
49. For example, Sale 2011.
50. Turvey et al. 2010.
51. See, for example, Roberts 2007; Jackson et al. 2011.
52. Nietzsche 1974, 91.
53. Blake 1977, 258.
54. Rendered after Hoban 1980, 6.

CHAPTER THREE

1. See, for example, Shaun Monson's (2010) documentary *Earthlings*. (Warning: Contains extremely violent content.)

2. Thomas 1983.
3. Bernstein 2005; Francis (Pope) 2015.
4. As the earliest use of "natural resources," the OED lists a 1776 passage from Adam Smith: *Inq. Wealth of Nations* II. v. iii. 562 "France, notwithstanding all its natural resources, languishes under an oppressive load of the same kind [national debt]." On the modern reconfiguration of nature as matter and mechanism, see Merchant 1980.
5. Sax 2011.
6. "A Special Report on the Sea," *Economist*, January 3 2009, 5–6; also see the press release of the Center for Biological Diversity, http://www .biologicaldiversity.org/news/press_releases/2015/deep-sea-mining-05-13 -2015.html; and Koslow 2007, chap. 8.
7. The common understanding of fishery is a place where fish are caught and the industry of catching and selling fish. More formal definitions render "fishery" as a hybrid concept, combining "the fish and fishers in a region, the latter fishing for similar or the same species with similar or the same gear types." See Blackhart et al., National Marine Fisheries Service, NOOA Fisheries Glossary, rev. ed., 2006, Department of Commerce, http://www .st.nmfs.noaa.gov/st4/documents/FishGlossary.pdf.
8. Danson 2011, 145.
9. Heidegger (1947) 2013, 278.
10. Quoted in Steiner 2011, 84.
11. Steiner 2011, 86.
12. Quoted in Steiner 2010, 130.
13. Thomas 1983, 51.
14. Heidegger 1977b, 129–30, emphasis added.
15. Steiner 2011, 83, 87.
16. See Lynas 2011; Ellis 2011; Brand 2009.
17. Nimmo 2011, 62.
18. Quoted in Jensen 2002, 36.
19. Cafaro 2015.
20. Steffen et al. 2007, 614.
21. See, for example, the *Economist*'s "Welcome to the Anthropocene," http:// www.economist.com/node/18744401.
22. Mander 2007.
23. Hamilton 2014.
24. Vidhi Doshi, "India Set to Start Massive Project to Divert Ganges and Brahmaputra Rivers," *Guardian*, 2016, https://www.theguardian.com/ global-development/2016/may/18/india-set-to-start-massive-project-to -divert-ganges-and-brahmaputra-rivers; "Robochop," *Economist*, 2013, http://www.economist.com/news/science-and-technology/21588049 -automated-jellyfish-exterminator-takes-sea-robochop.
25. Kaylee Heck, "Los Angeles Reservoir Covered with 96 Million Shade Balls to Conserve Water amidst Drought," ABC News, 2015, http://abcnews.go

.com/US/los-angeles-reservoir-covered-96-million-shade-balls/story?id=
33038319.

26. Kidner 2014.
27. Heidegger 1977a.
28. Heidegger 1977a.
29. Jonas 1974, 126.
30. The Invisible Committee 2009, 16, emphasis added.
31. Exceptions include Mumford 1967; Merchant 1980; and Thomas 1983. These works are multidimensional and comprehensive environmental histories.
32. Quoted in Steiner 2011, 88.

CHAPTER FOUR

1. Millennium Ecosystem Assessment 2005; Jackson 2008; Rockström et al. 2009; Butchart et al. 2010; Gibbs et al. 2010; Smil 2011; Steffen et al. 2011a, 2011b; Ellis et al. 2011a, 2012; Barnosky et al. 2011, 2012; Rees 2014; Butler 2015b; McKibben 2016; DellaSala et al. 2018.
2. Jones 2011.
3. Barnosky et al. 2012, 52. See also Ketcham's 2017b overview of the consequences of continued economic and population growth.
4. Barnosky et al. 2012, 55.
5. See, for example, Asafu-Adjaye et al. 2015.
6. Fuller 1982. Thanks to Dr. Robert Svoboda for calling my attention to Fuller's definition of a knot.
7. Butler 1988, 521.
8. Heidegger 1977b, 153.
9. Foley et al. 2005.
10. Ellis et al. 2013, 7978.
11. Derrida 1987, 131.
12. Hopkins 1983.
13. Manning 2004.
14. Nibert 2012, 2013.
15. Crosby 2015.
16. Moore 2016b, 88.
17. Thomas 1983, 194, 197.
18. Mazoyer and Roudart 2006, 288.
19. Quoted in Waters 1972, 278.
20. Thomas 1983, 30.
21. Quoted in McLuhan 1971, 85.
22. Roberts 2007, 23–25, chap. 4.
23. Vickers with McClenachan 2011, 128, emphasis added.
24. House 1999, 59.

25. For detailed analyses of the Anthropocene discourse, see Crist 2013b; Hamilton et al. 2015; Moore 2016a; Haraway 2016.
26. The other three in Bonneuil's typology are the postnature narrative, the ecocatastrophist narrative, and the eco-Marxist narrative. Bonneuil 2015, 18.
27. Crutzen and Schwägerl 2011; also cited in Revkin 2011.
28. "A Man-Made World," *Economist*, May 28, 2011, 81–83.
29. "The Anthropocene," *New York Times*, editorial, 2011, http://www.nytimes .com/2011/02/28/opinion/28mon4.html.
30. Steffen et al. 2007, 615.
31. For example, after sketching the emergence of hominid tool-making, rudimentary weapons, control of fire, and subsequent shift to an omnivorous diet, Will Steffen and his coauthors inform that the human brain size grew three-fold, giving "humans the largest brain-to-body ratio of any animal on the Earth." In turn, this enabled the development of language, writing, accumulation of knowledge, and social learning. "This has ultimately led to a massive—and rapidly increasing—store of knowledge upon which humanity has eventually developed complex civilizations and continues to increase its power to manipulate the environment. No other species now on Earth or in Earth history comes anywhere near this capability." Steffen et al. 2011a, 846.
32. Williams 2007, 136.
33. "A Man-Made World," *Economist*, May 28, 2011, 81–83.
34. Vince 2011, 33.
35. Jan Zalasiewicz, quoted in Owen 2010.
36. Paul Crutzen, quoted in Kolbert 2011.
37. As Derrick Jensen puts it: "Of course members of this culture, who have named themselves with no shred of irony or humility *Homo sapiens* would, as they murder the planet, declare this the age of man" (2013, 41).
38. Haraway 2016, 49.
39. Patel 2011, 101.
40. Marris 2009.
41. Pearce 2013. Novelty is not nature's norm any more than stasis is—a point I will return to in the last chapter. Also, understating the importance of *the speed of change* that humans are effecting (as this passage does) is misleading. Most changes in nature are not catastrophically rapid, which is why biodiversity tends to build over geological time, with the emergence of new species slightly outpacing background extinction.
42. "A Man-Made World," *Economist*, May 28, 2011, 81–83.
43. Steffen et al. 2011b, 740.
44. Kotchen and Young 2007.
45. Steffen et al. 2011a, 857–58.
46. Kirschenmann 2008.
47. Austin 1962.
48. Williams et al. 2015, 11.

49. Other coined terms in the Anthropocene literature do similar work of naturalizing the human impact. Examples include "sociocultural niche construction," "neobiota," "novel ecosystems," and "ecological engineering," all of which represent human activities as ecological phenomena. See, for example, Williams et al. 2015.
50. Berry 2002, 146.
51. Martin 1967.
52. Grayson and Meltzer 2003; Koch and Barnosky 2006; Barnosky 2008; Boulanger and Lyman 2014; Sandom et al. 2014.
53. Koch and Barnosky 2006; Barnosky 2008.
54. Koch and Barnosky 2006, 241.
55. Grayson and Meltzer 2003.
56. Wroe et al. 2004, 321.
57. Asafu-Adjaye et al. 2015.
58. Wroe et al. 2004, 321.
59. Dirzo and Raven 2003, 154.
60. Braje and Erlandson 2013, 20.
61. Wright 2005, 37.
62. Wright 2005, 63.
63. Wroe et al. 2004, 300. See also Grayson and Meltzer 2003.
64. Punke 2007, 73.
65. Moore 2016b.
66. The Great Acceleration refers to the period of human impact after World War II, and the initiation of post–Bretton Woods economic globalization. As Will Steffen and his coauthors state, "although the imprint of human activity on the global environment was, by the mid-twentieth century, clearly discernible beyond the pattern of Holocene variability in several important ways, the rate at which that imprint [has been growing] increased sharply at midcentury." Steffen et al. 2001a, 849.
67. Butler 1988, 520.
68. Butler 1988, 520.
69. Butler 1988, 522. See also Kidner 2014.
70. Butler 1999, 343.
71. Rodman 1980, 74.
72. Martha Nussbaum, quoted in Steiner 2011, 101.
73. Calarco 2012, 56.
74. Calarco 2012, 47.

CHAPTER FIVE

1. *Economist*, May 28, 2011, 11. See also Wohl 2013.
2. Guha 1989; Denevan 1992; Cronon 1995. See also Callicott and Nelson 1998, 2008, for classic and contemporary papers debating the idea of wilderness.

3. Keim 2014.
4. Kidner 2014, 473.
5. Quoted in Adam Vaughan, "Humans Have Destroyed a Tenth of Earth's Wilderness in 25 Years—Study," *Guardian*, 2016, https://www.theguardian.com/environment/2016/sep/08/humans-have-destroyed-a-tenth-of-earths-wilderness-in-25-years-study.
6. Soulé and Lease 1995; Caro et al. 2011.
7. Dinerstein et al. 2017.
8. Kingsnorth 2013.
9. Soulé and Noss 1998, 21.
10. "Think Big" 2011, 131.
11. Rolston 2015, 34.
12. Noss 2013.
13. Keim 2014.
14. Soulé and Noss 1998; Foreman 2004; Monbiot 2013; Wuerthner, Crist, and Butler 2014, 2015; Bekoff 2014; Johns 2016.
15. Ehrlich and Pringle 2008, 11583.
16. Foreman 2004; Monbiot 2013; Bekoff 2014.
17. Hettinger 2014.
18. Quoted in Johns 2016, 5.
19. Johns 2016, 4.
20. Pimm 2013, 57.
21. Whitty 2007, 89.
22. "Think Big" 2011, 131.
23. Hoekstra et al. 2005.
24. See Chew 2001 for an environmental history of the region.
25. Kidd 2012.
26. MacKinnon 2006; Kidd 2012.
27. Hopkins 1983.
28. Thomas 1983, 25.
29. Nash 2012, 302.
30. Ellis 2013a.
31. Mumford 1967, 186.
32. Denevan 1992.
33. Marris 2013.
34. Manning, personal communication.
35. Kareiva et al. 2011.
36. Denevan 1992; Mann 2002.
37. Marris 2013.
38. National Research Council 1970.
39. Hiss 2014.
40. Annie B. White, "Gray Wolf Conservation: A History of Wild Wolves in the United States," http://www.graywolfconservation.com/Wild_Wolves/history.htm.

41. "The Distribution of Mountain Lions: Distribution in the Americas," Balanced Ecology, Inc., http://www.balancedecology.org/MountainLionWebSite/Mountain_Lion_Distribution.html; "Mountain Lion," National Park Service, https://www.nps.gov/lake/learn/nature/mountainlion.htm.

42. "American Elk," Fish and Wildlife Habitat Management Leaflet 11:1–8, USDA Natural Resources Conservation Service, 1999, http://www.nrcs.usda.gov/Internet/FSE_DOCUMENTS/nrcs143_010000.pdf.

43. "Pronghorn," National Park Service, Yellowstone, https://www.nps.gov/yell/learn/nature/pronghorn.htm.

44. "Grizzly Bears: Basic Facts about Grizzly Bears," Defenders of Wildlife, https://www.defenders.org/grizzly bear/basic-facts; George Wuerthner "Habitat Conservation, Not Hunting, Saves Grizzly Bears," Wilderutopia, 2013, http://www.wilderutopia.com/environment/wildlife/george-wuerthner-habitat-conservation-saves-grizzly-bears-not-hunters/.

45. Wilcove 2008a, 109.

46. "Woodland Caribou: Basic Facts about Woodland Caribou," Defenders of Wildlife, http://www.defenders.org/woodland-caribou/basic-facts; "Woodland Caribou (*Rangifer tarandus caribou*)," US Fish and Wildlife, Environmental Conservation Online Services, https://ecos.fws.gov/ecp0/profile/speciesProfile?spcode=A088.

47. Leopold (1949) 1968, 111.

48. "Passenger Pigeon," Museum of Vertebrate Zoology at Berkeley, http://www.si.edu/Encyclopedia_SI/nmnh/passpig.htm; Barry Yeoman, "Why the Passenger Pigeon Went Extinct," *Audubon Magazine*, 2014, http://mvz.berkeley.edu/Newsletter/newsletter_files/201408/Passenger_Pigeon.php.

49. Baker and Hill 2003, 288.

50. Danson 2011, 93.

51. Mowat (1984) 2004.

52. Mann 2002.

53. Ripple et al. 2015, 7.

54. Curry 2012, 142–43.

55. LaDuke 1997, 26.

56. Metzner 1993.

57. Graham Harvey, quoted in Curry 2012, 143. See also Kimmerer 2003, 103.

58. LaDuke 1997, 25.

59. LaDuke 1997, 26.

60. Mills 2010, 9. See also House 1999.

61. For example, early on white settlers roped many native peoples into the fur trade. See Valliant 2005.

62. Marshall (1930) 1998, 86.

63. Mann (2005) 2011, 283.

64. Baker 2005, emphasis added.

65. Baker 2005.

66. Curry 2012, 176.
67. Steinberg 2002, 20.
68. Kimmerer 2003, 141.
69. See, for example, Kimmerer 2010.
70. Quoted in Keim 2014.
71. Mann 2002.
72. Mann (2005) 2011, 286.
73. Mann 2002.
74. Turner 1996, 21.
75. Shabekoff 2003, 9.
76. Thomas 1983, 18.
77. Bruce and William Catton, quoted in Shabekoff 2003, 14.
78. "Demographic History of the United States," Wikipedia, https://en
    .wikipedia.org/wiki/Demographic_history_of_the_United_States. See also
    Steinberg 2002, 62.
79. Slotkin 1973, 9.
80. Slotkin 1973, 17.
81. Quoted in Nye 2003.
82. Thomas 1983, 194.
83. Shabekoff 2003, 31–33.
84. Steinberg 2002, 60.
85. Rodman 1980, 55.
86. Nye 2003.
87. Nye 2003.
88. Nye 2003.
89. Manning 2013.
90. Klare 2012.
91. Miller 2017.
92. Evans 2015.
93. Thoreau 1992, 42.
94. Nietzsche (1954) 1968, 129.
95. William Wordsworth, quoted in Thomas 1983, 267.

CHAPTER SIX

1. According to ecologist David Wilcove, all seven species of sea turtles are
   endangered. Wilcove 2008a, 150.
2. Wilcove 2008a, 160.
3. Safina 2011, 15.
4. Jackson and Alexander 2011, 5.
5. Roberts 2007, chap. 5.
6. Sala and Jackson 2011, 196.
7. Wilcove 2008a, 159–60.
8. Snyder 1990, 5.

9. Eckersley 2005, 372, 378. See also Nash 1989; Rodman 1987.
10. Nash 2012, 305.
11. See *Merriam-Webster*, online dictionary, s.v., "freedom," http://www
    .merriam-webster.com/dictionary/freedom.
12. Chakrabarty 2009, 208.
13. Rodman 1980, 62.
14. Mumford 1964, 6.
15. Mumford 1964, 6.
16. Weber (1905) 2013.
17. Balabanian 2006.
18. Jenkins 2003, 1176.
19. "Aker BioMarine Secures Sustainable Future for Krill in the Antarctic
    Southern Ocean," Marine Stewardship Council, 2015, https://www.msc
    .org/newsroom/news/aker-biomarine-secures-sustainable-future-for-krill
    -in-the-antarctic-southern-ocean.
20. "A Special Report on the Sea," *Economist*, 2008, 13. It is usually after
    "stocks" become "overexploited" that attempts to manage them and cur-
    tail overfishing set in.
21. Danson 2011, 116.
22. Sumaila and Pauly 2011, 28. See also Roberts 2007, chap. 22.
23. Jeremy Jackson, presentation, 2014, https://www.youtube.com/watch?v=
    NyWHSflfiag.
24. Quoted in the documentary *End of the Line* (2009).
25. For a historical account of *Mare Liberum*, see Vidas 2011.
26. Jackson 2008, 11463.
27. Pauly and Zeller 2016.
28. "A Special Report on the Sea," *Economist*, 2008, 14.
29. As available fish for commercial fishing decline or nosedive, fishermen go
    deeper into debt in order to invest in ever more efficient technologies and
    gear. Not only do the better technologies further devastate the remaining
    fish, but fishermen, now heavily in debt, press for fewer restrictions on the
    fishing so they can pay back their loans. See Smith 2002–3. For detailed
    descriptions of institutional frameworks related to the exploitation of the
    ocean, see Danson 2011; Erle 2009.
30. Smith et al. 2010.
31. Feltman 2015.
32. "Tern Limits," *Economist*, August 31, 2013. See also Pew Environment
    Group, "Out of Balance: Industrial Fishing and the Threat to Our Ocean,"
    http://www.pewtrusts.org/en/research-and-analysis/reports/2008/05/21/
    out-of-balance-industrial-fishing-and-the-threat-to-our-ocean.
33. Danson 2011, 144.
34. Fulmer 2009, 30, 31.
35. US National Research Council, quoted in https://en.wikipedia.org/wiki/
    Infrastructure#cite_note-8 .

36. Quoted in Fulmer 2009, 30.
37. Noss 1998.
38. Kroll 2015, 6.
39. Soron 2008, 111.
40. On the ecological impact of roads, see also Sanderson et al. 2002, 893–94; Coffin 2007; Daigle 2010; van der Ree et al. 2011.
41. "Earth on Edge," interview with Adrian Forsyth, Bill Moyers Reports, http://www-tc.pbs.org/earthonedge/program/forsyth.pdf.
42. "Roads & Off Road Vehicles," Forest Protection and Restoration, Oregon Wild, http://www.oregonwild.org/forests/forest-protection-and -restoration/roads.
43. "The Transamazonian Highway," *Mongabay*, http://rainforests.mongabay .com/08highway.htm.
44. Laurance 2015.
45. R. Butler 2013.
46. Laurance 2015.
47. Trombulak and Frissell 2000.
48. Dulac 2013.
49. John Dulac, personal communication.
50. "Public-Private Partnerships in Roads," PPPIRC (Public-Private Partnership in Infrastructure Research Center), World Bank Group, https://ppp .worldbank.org/public-private-partnership/sector/transportation/roads -tolls-bridges/road-concessions.
51. Noss 1998.
52. Laurance goes on to note that "Brazil's Balbina Dam flooded 240,000 hectares of rainforest—an area larger than Buenos Aires. Plans are afoot to build another 150 major dams in the Amazon—each of which will flood expanses of forest and require networks of new roads for dam and power line construction." Laurance 2015.
53. Alexander 2014, 7.
54. Alexander 2014, 7.
55. Andrew Browne, "U.S. Cedes Clout to China under Massive Infrastructure Plan," *Wall Street Journal*, January 31, 2018.
56. Blackwelder 2015.
57. Rodman 1980, 62.
58. Mumford 1964, 5.
59. Kareiva and Marvier 2012, 968.
60. Lulka 2004, 443.
61. Roberts 2007, 44.
62. Wilcove 2008a, 168.
63. Wilcove and Wikelski 2008, 1361.
64. Wilcove 2008b.
65. Wilcove 2008b.
66. Wilcove and Wikelski 2008, 1361.

67. Lester et al. 2016; Wilcove 2008a, 5.
68. Wilcove 2008a, 201; Wilcove 2008b.
69. Wilcove 2008b.
70. "Swainson's Thrush Life History," All about Birds, Cornell Lab of Ornithology, https://www.allaboutbirds.org/guide/Swainsons_Thrush/lifehistory.
71. Hill 2015; Wilcove 2008a, 29.
72. Wilcove 2008a, 46.
73. Waldman 2010.
74. Wilcove 2008a, 184–85.
75. Wilcove 2008b.
76. Wilcove and Wikelski 2008, 1361.
77. Hill 2015; see also "Humpback Whale (*Megaptera novaeangliae*)," NOAA (National Oceanic and Atmospheric Administration) Fisheries, www.nmfs.noaa.gov/pr/species/mammals/whales/humpback-whale.html.
78. Wilcove 2008a, 138.
79. Wilcove 2008a, 103.
80. See Engelman 2016; Laurance 2015; Laurence et al. 2015.
81. Bauer et al. 2015.
82. Wilcove 2008a, 96–87.
83. Pearce 2011.
84. Ceballos et al. 2015, 27.
85. Gates et al. 2010.
86. Freese 2015.
87. Freese 2015.
88. Wilcove 2008a, 99, emphasis added.
89. Lulka 2004.
90. Wilcove 2008a, 120. The curtailment of bison by any means necessary is advocated especially by ranchers who fear the transmission of the disease brucellosis to their cattle. Yet implementing techniques and policies to insulate the cattle from contact—instead of killing the bison—is not entertained as an alternative by the human establishment.
91. Wilcove 2008a, 95.
92. Lulka 2004, 449.
93. Lulka 2004, 459–60.
94. Butler 2015a, xxiii.
95. Birch 1995.
96. Birch 1995, 344.
97. Quoted in McLuhan 1971, 71.
98. Birch 1995, 352.
99. Birch 1995, 351.
100. Berry 1999.
101. Naess 1995, 224.
102. Turner 1996, 68.

103. Marcuse 1969, 29.
104. Thoreau 1991, 104; Thoreau 1992, 14; Brower 2000.

CHAPTER SEVEN

1. Gerland et al. 2014.
2. For example, sustainability thinker William Rees maintains that "the overarching problem is one that the mainstream has yet to acknowledge: on a planet already in overshoot, there is *no* possibility of raising even the *present* world population to developed country material standards sustainably with known technologies and available resources." Rees 2014, 3, emphasis in original.
3. See, for example, Foley et al. 2011.
4. Biello 2009b.
5. Invisible Committee 2009, 14.
6. Reid 2012.
7. Patel 2011, 50.
8. Whitty 2011.
9. Barnosky 2012.
10. For example, *Nature* reporter Jason Clay writes about the need "to reform the global food system by increasing food production without damaging biodiversity." Clay 2011, 287.
11. Baulcombe et al. 2009.
12. Gibbs et al. 2010.
13. Rees 2014.
14. Adam Vaughan, "Humans Have Destroyed a Tenth of Earth's Wilderness in 25 Years—Study," *Guardian*, 2016, https://www.theguardian.com/environment/2016/sep/08/humans-have-destroyed-a-tenth-of-earths-wilderness-in-25-years-study.
15. Ehrlich and Ehrlich 2004, 183.
16. Hansen and Gale 2014.
17. Maverick 2014.
18. Vijay et al. 2016.
19. Ellis 2013b.
20. Quoted in Federoff and Brown 2010, 345.
21. Quoted in Weisman 2013, 424–25.
22. Palmer 2015.
23. Duncan 2013, 13.
24. Weisman 2013, 58.
25. The fact that people who developed Green Revolution technologies had the good intention of saving lives does not hold us to perpetuating a food system that after decades has clearly revealed its inherent ecological destructiveness.
26. Marx 1996.

27. Weis 2013.
28. Citing India's National Crime Records Bureau, Alan Weisman reports that 270,000 Indian farmers have taken their own lives since 1995 (2013, 336). See also "India's Farmer Suicides: Are Deaths Linked to GM Cotton?—in Pictures," *Guardian*, https://www.theguardian.com/global-development/gallery/2014/may/05/india-cotton-suicides-farmer-deaths-gm-seeds.
29. Jackson 2011.
30. See Manning's overview of the history of agriculture (2004).
31. Manning 2004, 2016.
32. Keith 2011, 208.
33. In recent decades, the productivity of farm animals (to grow faster, have larger litters, produce more milk, and so on) has been increased via "breeding techniques" equivalent to experimentation on sentient beings—in the worst sense of the word "experimentation." See Michael Moss, "Livestock Pay Price in Industry Research," *New York Times*, January 20, 2015.
34. Scully 2003; Tuttle 2005; Dawkins and Bonney 2008; Imhoff 2010.
35. "Feeding the World" (special issue), *Economist*, 2011, http://www.burness.com/economist-special-report-feeding-world/.
36. The feed is typically laced with antibiotics to make the animals grow faster and preempt the diseases that come from crowding. The livestock industry is the greatest consumer of antibiotics, contributing to the emergence of antibiotic resistant strains of pathogens. See Foer 2009.
37. Weisman 2013.
38. Pollan 2014.
39. CIMMYT to Expand in India, October 2011 BGRI E-Newsletter.
40. Cotton being inedible, while corn and soybeans are mostly fed to confined animals and used in automobile fuel tanks.
41. Quoted in Bourne 2009.
42. Carson (1962) 1994, 12.
43. Citing recent research, Raj Patel writes that "sustainable organic farming could sequester up to 40 percent of current CO_2 emissions." Patel 2011, 163.

CHAPTER EIGHT

1. Speth 51.
2. Magdoff and Foster 2011, 108.
3. Campbell 2007, 240.
4. Curry 2012, 253.
5. "The State of Consumption Today," Worldwatch Institute, 2018, http://www.worldwatch.org/node/810.
6. McKee and Chambers 2011, 54. See also McKee, Chambers, and Guseman 2013.

7. Weisman 2013, 246.
8. Weisman 2013, 248. Scott Moore, "Pakistan's Water Crisis Is a Ticking Bomb," *National Interest*, 2018, http://nationalinterest.org/feature/pakistans-water-crisis-ticking-time-bomb-24347.
9. McKibben 1998, 174–75.
10. McKee and Chambers 2011, 55.
11. McKibben 2016.
12. Halweil 2006; "Can Organic Food Feed the World?" 2015; Reganold and Wachter 2016.
13. Smil 1997.
14. Pimentel et al. 2010, 601.
15. Georgescu-Roegen 1975.
16. Whitty 2010.
17. Ripple et al 2015, 7.
18. Magdoff and Foster 2011, 32.
19. Ravallion 2009; Kharas 2017.
20. Kharas 2017.
21. Hayden 2010, 52.
22. Engelman 2016.
23. Durning 1992; Dauvergne 2008.
24. Engelman 2012a, 121.
25. Engelman 2012a, 122.
26. Smail 1997, 234.
27. Potts 2009, 3122.
28. See Weeden and Palomba 2012.
29. Chesler 2010, 312.
30. Goldberg 2009.
31. Potts 2009.
32. Campbell 2007, 242.
33. Potts 2009, 3120.
34. See Campbell and Bedford 2009.
35. Engelman 2016.
36. Ottaway et al. 2007, 32.
37. Foer 2009, 262.
38. Monbiot 2015; Andersen and Kuhn 2015.
39. Quoted in Foer 2009, 238.
40. George Monbiot, "I've Converted to Veganism to Reduce My Impact on the Living World," *Guardian*, 2016, https://www.theguardian.com/commentisfree/2016/aug/09/vegan-corrupt-food-system-meat-dairy.
41. See Freeman 2007; Schlosser 2011.
42. Machovina 2015, 427.
43. On the silence surrounding the population question during the 1990s and 2000s, see Campbell 2007.
44. Daily and Ehrlich 1994. See also John Vidal, "Cut World Population and Redistribute Resources, Expert [Paul Ehrlich] Urges," *Guardian*,

2012, https://www.theguardian.com/environment/2012/apr/26/world
-population-resources-paul-ehrlich. Weisman 2013, 92–94.

45. Pimentel et al. 2010.
46. Engelman 2008, 205.
47. Fabic et al. 2015, 1928–29.
48. Bongaarts 2016, 412.
49. Fabic et al. 2015, 1930.
50. Fabic et al. 2015, 1929–30.
51. Campbell and Bedford 2009.
52. Campbell and Potts 2012.
53. Engelman 2016.
54. Engelman 2012a, 123.
55. Weisman 2013, 418.
56. Engelman 2016.
57. Campbell and Potts 2012.
58. Engelman 2016.
59. Kaidbey and Engelman 2017, 185.
60. Kaidbey and Engelman 2017, 185–86.
61. Kaidbey and Engelman 2017, 181, 187.
62. Kaidbey and Engelman 2017, 189.
63. Sedgh et al. 2014.
64. Engelman 2012b.
65. Engelman 2012a, 128.
66. See McKibben 1998; Weisman 2013, 415.
67. Weisman 2013, 181–82.
68. Haraway 2016, 103.
69. Pimentel et al. 2010, 606.
70. Weisman 2013, 240.
71. Ezeh et al. 2012.
72. Smeeding 2014.
73. Ehrlich and Ehrlich 2006, 47.
74. Ehrlich and Ehrlich 2006, 48–49.
75. Tal 2017, 51.
76. Plumwood 2008.
77. FAO 2015, 38.
78. FAO 2015, 38.
79. Haraway 2016, 100.
80. Ehrlich and Ehrlich 2013.
81. Weisman 2013, 393.
82. See http://www.economist.com/node/14744915.
83. MacKay 2001; Pretty 2007, 139–52.
84. Kirschenmann 2008.
85. Fukuoka (1978) 2009, 117.
86. Danson 2011, 116–17.
87. Berry 2009, xii.

CHAPTER NINE

1. Plato 1961, 1183.
2. Wu and Loucks 1995.
3. Krause 2012.
4. Krause 2012.
5. See, for example, Botkin 1990; Jelinski 2005.
6. In Bly 1980, 123.
7. See Lovelock 1988; Crist and Rinker 2009.
8. Lenton 2004.
9. See "Gopher Tortoise," Florida Fish and Wildlife Conservation Commission, http://myfwc.com/wildlifehabitats/managed/gopher-tortoise/; Gopher Tortoise Council, http://www.gophertortoisecouncil.org/.
10. Hiss 2014.
11. "Standing Tall: How Restoring Longleaf Pine Can Help Prepare the Southeast for Global Warming," National Wildlife Federation, 2009, https://www.nwf.org/pdf/Reports/LongleafPineReport.pdf; "Gopher Tortoise (Gopherus Polyphemus)," US Fish and Wildlife Service, 2012, https://www.fws.gov/daphne/Fact_Sheets/GopherTortoiseFS2012.pdf.
12. They are threatened by habitat loss and degradation, collection by people for pets or for eating them, predation by mesopredators and invasive red ants, herbicide use for "forest management," and climate-change driven changes in precipitation and sea levels.
13. Mathews 2011.
14. Whitty 2011, 176.
15. Quotes in Roberts 2007, chap. 9.
16. Westbroek 1991, chap. 6.
17. Whitty 2011, 208.
18. Roberts 2007, 305.
19. Watch the fascinating *Great White Shark* 3D movie. https://www.imax.com/movies/m/great-white-shark-3d/.
20. Safina 2011, 17.
21. Zimmer 2010.
22. Valliant 2005, 11.
23. Margulis, quoted in Rose 2013, 50.
24. Butler and Lubarsky 2017.
25. Jalaluddin Rumi, "Where Everything Is Music."
26. Whitehead 1966, 138.
27. *Nature* 476 (August 25, 2011): 375.
28. Margulis 1993, 279. See also Margulis 1998.
29. Rinker 2011.
30. Simard 2016.
31. Kimmerer 2003, 57.
32. Ripple et al. 2015, 6.

33. House 1990, 160.
34. Kimmerer 2003, 144.
35. Kimmerer 2003, 146.
36. In 2013 red knots were declared threatened. "The primary factor in the re-cent decline of the species was reduced food supplies in Delaware Bay due to commercial harvest of horseshoe crabs." See "Service Proposes to List Red Knot as a Threatened Species under the Endangered Species Act," US Fish and Wildlife Service, press release, 2013, https://www.fws.gov/news/ShowNews.cfm?ID=60042DE0-FB9E-C978-063157265CB076C1.
37. Roberts 2012.
38. Wilcove 2008a, 88.
39. Wilcove 2008a, 109–10.
40. Kimmerer 2010, 144.
41. Kingsnorth 2013.
42. Kohák 1987, 90.
43. Nietzsche, quoted in Heidegger 1977d, 85.
44. Locke 2013; Wilson 2016; Dinerstein et al. 2017.
45. Birch 1990, 350.
46. Brower 2000, 52.
47. Bennett 2011; Mander 2013; Pflanz 2014.
48. For more detailed discussion about protected areas, including debates sur-rounding "fortress conservation," see Crist 2015; Wuerthner et al. 2015.
49. See Mackey et al. 2015; Mills 2010; Wilson 2002, 161.
50. Baeten et al. 2010, 1447.
51. Baeten et al. 2010, 1450, 1451.
52. Todd Woody, "Sylvia Earle: Stop Fishing on the High Seas to Save the Ocean," interview with Sylvia Earle, *News Deeply*, 2017, https://www.newsdeeply.com/oceans/community/2017/06/14/sylvia-earle-stop-fishing-on-the-high-seas-to-save-the-ocean.
53. "Stop Fishing the High Seas, Say Scientists, for Climate and Ecology," *Ecologist*, 2014. See also "Interview with Captain Paul Watson," *Ecological Citizen*, 2018.
54. Noss et al. 2011; Locke 2013; Wilson 2016; Dinerstein et al. 2017.
55. Quoted in Hiss 2014.
56. Ripple et al. 2015, 8.
57. Kopnina 2012. See also Crist 2015 and references therein.
58. Kopnina 2012, 18.
59. Tompkins and Butler 2016.
60. Wilson 2006, 89.
61. Thoreau 1991, 255.
62. Thoreau 1991, 255.
63. Mills 1997, 277.
64. Nash 2012.
65. Gurdjieff, quoted in Ouspensky (1950) 1977, 357.

66. Kirschenmann 2017.
67. Sale 1997, 233.
68. See Taylor 2000 for an overview.
69. Quoted in McLuhan 1971, 54.
70. McLuhan 1971, 119.
71. Quoted in Stromberg 1975, 33.
72. See Weber (1905) 2013; Schutz 1999.
73. Kloppenburg et al. 1996, 37.
74. John Davis, quoted in Taylor 2000, 57.
75. Taylor 2000, 50.
76. Sale 1997, 220.
77. Rose 2014.
78. Peter Berg and Raymond Dasmann, quoted in House 1999, 126.
79. Plumwood 2008.
80. See, for example, Patel 2009.
81. Trombulak 2011, 135.
82. Taylor 2000.
83. Orr 2004.
84. Quoted in McLuhan 1971, 6.
85. Quoted in House 1999, 64.
86. Paul Shepard, quoted in House 1999, 158.
87. Patel 2012.
88. Nash 2012.
89. Drengson 2004.
90. Taylor 2000, 51.
91. Mumford 1964, 1–2.
92. Benhabib 2004, 117.
93. Benhabib 2004, 117.
94. Rorty (1991) 1996.
95. Karl Marx, quoted in Osborn 2010, 126.
96. Steiner 2011.
97. Haraway 2008.
98. Quoted in Taylor 2000, 50.
99. Berry 2002, 145.
100. Rose 2013, 62.
101. Plotica 2016, 485.
102. Quoted in Taylor 2000, 59.
103. Petrini 2001; Andrews 2008; Berardi 2009.
104. Heidegger 1975, 190.
105. Heidegger 1977a.
106. We can see something "now as one thing now as another.—So we interpret it, and *see* it as we *interpret* it." Wittgenstein (1953) 1968, 193, emphasis in original.
107. Wittgenstein (1953) 1968, 48.

108. Coetzee 2000.

109. Bane 2007; Griscom et al. 2017.

110. Dan Deudney, quoted in Taylor 2000, 63.

111. See "Elon Musk Reveals His Plan for Colonizing Mars," 2016, https://www
.youtube.com/watch?v=W9olSzNOh8s.

EPILOGUE

1. Jonathan Foley, quoted in Palmer 2015.

2. "World Nuclear Weapon Stockpile," Ploughshares Fund, http://www
.ploughshares.org/world-nuclear-stockpile-report; "Stockpiles of Nuclear
Weapons around the World—in Data," *Guardian*, 2017, https://www
.theguardian.com/world/2017/mar/11/stockpiles-of-nuclear-weapons
-around-the-world-in-data.

3. Patel 2011, 79.

4. Jonathan Watts and John Vidal, "Environmental Defenders Being Killed
in Record Numbers Globally, New Research Reveals," *Guardian*, 2017,
https://www.theguardian.com/environment/2017/jul/13/environmental
-defenders-being-killed-in-record-numbers-globally new-research-reveals.

5. Marcuse 1969, 24–25.

6. Marcuse 1969, 19.

7. Marcuse 1969, 34.

8. Johns 2002, 13.

9. Turner 1996, 57.

10. Turner 1996, 58.

11. Rodman 1980, 59.

12. Steiner 2011, 114. See also Henning 2016.

13. Heidegger 1977c, 39. The full passage reads: "But the surmounting of
a destining of Being—here and now, the surmounting of Enframing—
. . . comes to pass out of the arrival of another destining, a destining
that does not allow itself to be logically or historiographically predicted
or to be metaphysically construed as a sequence belonging to a process
of history." In plainer English this passage can be interpreted as stating
that the conception and treatment of the world as "standing reserve" will
be surmounted as Earth's and humanity's fated destination, through a
veering away from that seemingly set historical course toward an entirely
new destination by means of "the switchman" (to borrow Weber's classic
metaphor for historical change) of a profound shift in human consciousness in relationship with the biosphere. While such a consciousness shift
may seem difficult to imagine, it is far from an impossibility and, as Heidegger incisively noted, how and by what means it might occur cannot be
predicted. The shift will be adumbrated by surprising events; for example,
Pope Francis's 2015 Encyclical may qualify as one.

References

Agamben, Giorgio. 2004. *The Open: Man and Animal.* Stanford, CA: Stanford University Press.

Alexander, Nancy. 2014. "The Emerging Multi-Polar World Order: Its Unprecedented Consensus on a New Model for Financing Infrastructure Investment and Development." Washington, DC: Heinrich Böll Foundation.

Andersen, Kip, and Keegan Kuhn. 2015. *Cowspiracy: The Sustainability Secret.* Documentary. A.U.M. Films and First Spark Media.

Andrews, Geoff. 2008. *The Slow Food Story: Politics and Pleasure.* Montreal: McGill-Queen's University Press.

Asafu-Adjaye, John, et al. 2015. "An Ecomodernist Manifesto." http://www.ecomodernism.org/.

Austin, J. L. 1962. *How to Do Things with Words.* Cambridge, MA: Harvard University Press.

Baeten, Lander, et al. 2010. "Unexpected Understorey Community Development after 30 Years in Ancient and Post-Agricultural Forests." *Journal of Ecology* 98:1447–53.

Bagheera. N.d. "The Heath Hen: Extinct." http://www.bagheera.com/inthewild/ext_heathhen.htm.

Baker, Bruce, and Edward Hill. 2003. "Beaver (*Castor Canadensis*)." In *Wild Mammals of North America: Biology, Management, and Conservation*, edited by G. A. Feldman et al., 288–310. Baltimore, MD: Johns Hopkins University Press.

Baker, Kevin. 2005. "*1491*: Vanished Americans." *New York Times*, October 9.

Balabanian, Norman. 2006. "On the Presumed Neutrality of Technology." *IEEE Technology and Society Magazine*, Winter, 15–25.

Bane, Peter. 2007. "Storing Carbon in Soil: The Possibilities of a New American Agriculture." *Permacultural Activist* 65:56–57.

Barnosky, Anthony. 2008. "Megafauna Biomass Tradeoff as a Driver of Quaternary and Future Extinctions." *PNAS* 105:11543–48.

Barnosky, Anthony, et al. 2012. "Approaching a State Shift in Earth's Biosphere." *Nature* 486:52–58.

Barnosky, Anthony, et al. 2011. "Has the Earth's Sixth Mass Extinction Already Arrived?" *Nature* 471:51–57.

Barr, Bradley, and James Lindholm. 2002–3. "Conserving the Sea: Using Lessons from the Land." *Wild Earth* 12 (4): 56.

Bauer, Hans, et al. 2015. "Lion (*Panthera Leo*) Populations Are Declining Rapidly across Africa, Except in Intensively Managed Areas." *PNAS* 112 (48): 14894–99.

Bauer, S., and B. J. Hoye. 2014. "Migratory Animals Couple Biodiversity and Ecosystem Functioning Worldwide." *Science* 344:54–58.

Baulcombe, Sir David, et al. 2009. "Reaping the Benefits: Science and the Sustainable Intensification of Global Agriculture." London: Royal Society.

Bekoff, Marc. 2014. *Rewilding Our Hearts: Building Pathways of Compassion and Coexistence*. Novato, CA: New World Library.

Benhabib, Seyla. 2004. "Reclaiming Universalism: Negotiating Republican Self-Determination and Cosmopolitan Norms." Presented at the Tanner Lectures on Human Values, University of California, Berkeley, March 15–19.

Bennett, Elizabeth. 2011. "Another Inconvenient Truth: The Failure of Enforcement Systems to Save Charismatic Species." *Oryx* 45 (4): 476–79.

Berardi, Franco "Bifo." 2009. *The Soul at Work: From Alienation to Autonomy*. Los Angeles: Semiotext(e).

Bernstein, Ellen. 2005. *Splendor of Creation: A Biblical Ecology*. Pilgrim.

Berry, Thomas. 1999. *The Great Work: Our Way into the Future*. New York: Three Rivers Press.

Berry, Wendell. 2002. *Citizenship Papers*. Berkeley, CA: Counterpoint.

Berry, Wendell. 2009. Preface to *The One-Straw Revolution*, by Masanobu Fukuoka, xi–xv. New York Review Books.

Biello, David. 2009a. "Another Inconvenient Truth: The World's Growing Population Poses a Malthusian Dilemma." *Scientific American*, October 2. http://www.scientificamerican.com/article/growing-population-poses-malthusian-dilemma/.

Biello, David. 2009b. "The Dam Building Boom: Right Path to Clean Energy?" *Yale Environment 360*. http://e360.yale.edu/feature/the_dam_building_boom_right_path_to_clean_energy/2119/.

Birch, Thomas. 1995. "The Incarceration of Wildness: Wilderness Areas as Prisons." In *Deep Ecology for the 21st Century*, edited by G. Sessions, 339–55. Boston: Shambhala.

Bittman, Mark. 2008. "Rethinking the Meat-Guzzler." *New York Times*, January 27.

Blackwelder, Brent. 2015. "Massive Global Infrastructure Projects Could Prevent Achievement of a Sustainable Economy While Undermining

Life Support Systems of the Earth." *Daly News*, January 8. www.steadystate
.org.

Blake, William. 1977. *The Portable Blake*. London: Penguin Classics.

Bly, Robert, ed. 1980. *News of the Universe: Poems of Two-Fold Consciousness*. San
Francisco: Sierra Club Books.

Boddice, Rob ed. 2011. *Anthropocentrism: Humans, Animals, Environment*.
Leiden: Brill.

Böhm, Monika. 2013. "The Conservation Status of the World's Reptiles." *Biological Conservation* 157:372–85.

Bongaarts, John. 2016. "Slow Down Population Growth." *Nature* 530:409–12.

Bonneuil, Christophe. 2015. "The Geological Turn: Narratives of the Anthropocene." In *The Anthropocene and the Global Environmental Crisis*, edited by
C. Hamilton, C. Bonneuil, and F. Gemenne, 17–31. London and New York:
Routledge.

Botkin, Daniel. 1990. *Discordant Harmonies: A New Ecology for the 21st Century*.
Oxford: Oxford University Press.

Boulanger, Matthew, and R. Lee Lyman (2014). "Northeastern North American Pleistocene Megafauna Chronologically Overlapped Minimally with
Paleoindians." *Quaternary Science Reviews* 85:35–46.

Bourne, Joel. 2009. "The Global Food Crisis: The End of Plenty." *National Geographic Magazine*, June.

Braje, Todd, and Jon Erlandson. 2013. "Human Acceleration of Plant and
Animal Extinctions: A Late Pleistocene, Holocene, and Anthropocene
Continuum." *Anthropocene* 4:14–23.

Brand, Stewart. 2009. *Whole Earth Discipline: An Ecopragmatist Manifesto*. London: Viking.

Brower, David, with Steve Chapple. 2000. *Let the Mountains Talk, Let the Rivers
Run*. Gabriola Island, BC: New Society Publishers.

Buber, Martin. (1970) 1996. *I and Thou*. New York: Touchstone.

Butchart, Stuart et al. 2010. "Global Biodiversity: Indicators of Recent Declines." *Science* 328:1164–68.

Butler, Judith. 1988. "Performative Acts and Gender Constitution: An Essay in
Phenomenology and Feminist Theory." *Theatre Journal* 40 (4): 519–31.

Butler, Judith. 1999. "Subjects of Sex/Gender/Desire." In *The Cultural Studies
Reader*, edited by S. During, 340–53. London and New York: Routledge.

Butler, Rhett. 2008. "Destruction of Wetlands Worsens Global Warming."
Mongabay. https://news.mongabay.com/2008/07/destruction-of-wetlands
-worsens-global-warming/.

Butler, Rhett. 2013. "Deforestation in the Congo." *Mongabay*. http://rainforests
.mongabay.com/congo/deforestation.html.

Butler, Tom. 2015a. "Introduction: Protected Areas and the Long Arc toward
Justice." In *Protecting the Wild: Parks and Wilderness, the Foundation for
Conservation*, edited by G. Wuerthner, E. Crist, and T. Butler, xx–xxvii.
Washington, DC: Island Press.

Butler, Tom, ed. 2015b. *Overdevelopment, Overpopulation, Overshoot.* Goff Books.

Butler, Tom, and Sandra Lubarsky. 2017. *For Beauty: Douglas R. Tompkins—Aesthetics and Activism.* Berkeley, CA: David Brower Center.

Cafaro, Philip. 2015. "Three Ways to Think about the Sixth Mass Extinction." *Biological Conservation* 192:387–93.

Cafaro, Philip, and Eileen Crist, eds. 2012. *Life on the Brink: Environmentalists Confront Overpopulation.* Atlanta: University of Georgia Press.

Calarco, Matthew. 2012. "Identity, Difference, Indistinction." *New Centennial Review* 11:41–60.

Calarco, Matthew. 2014. "Being Toward Meat: Anthropocentrism, Indistinction, and Veganism." *Dialectical Anthropology* 38 (4): 415–29.

Callicott, Baird, and Michael Nelson, eds. 1998. *The Great New Wilderness Debate.* Atlanta: University of Georgia Press.

Callicott, Baird, and Michael Nelson, eds. 2008. *The Wilderness Debate Rages On.* Atlanta: University of Georgia Press.

Campbell, Joseph. 1964. *Occidental Mythology: The Masks of God.* New York: Penguin.

Campbell, Martha. 2007. "Why the Silence on Population?" *Population & Environment* 28:237–46.

Campbell, Martha, and Karen Bedford. 2009. "The Theoretical and Political Framing of the Population Factor in Development." *Philosophical Transactions of the Royal Society B* 364:3101–3.

Campbell, Martha, and Malcolm Potts. 2012. "Do Economists Have Frequent Sex?" *Population Press*, December 20. https://populationpress.org/2012/12/20/do-economists-have-frequent-sex-by-martha-campbell-and-malcolm-potts/.

"Can Organic Food Feed the World?" 2015. *Wall Street Journal*, July 12. https://www.wsj.com/articles/can-organic-food-feed-the-world-1436757046.

Caro, Tim et al. 2011. "Conservation in the Anthropocene." *Conservation Biology* 26 (1): 185–88.

Carson, Rachel. (1962) 1994. *Silent Spring.* Boston: Houghton Mifflin Company.

Casey, Michael. 2015. "Seabirds Suffering Massive Population Declines." CBS News, July 14. http://www.cbsnews.com/news/seabirds-suffering-massive-population-declines/.

Casey, Susan. 2010. "Garbage In, Garbage Out." *Conservation Magazine*, January 14.

Cavalieri, Paola. 2006. "The Animal Question: A Reexamination." In *In Defense of Animals: The Second Wave*, edited by P. Singer, 54–68. Malden, MA: Blackwell.

Ceballos, Gerardo, Anne, Ehrlich, and Paul Ehrlich. 2015. *The Annihilation of Nature: Human Extinction of Birds and Mammals.* Baltimore, MD: Johns Hopkins University Press.

Ceballos, Gerardo, Andrés García, and Paul Ehrlich. 2010. "The Sixth Extinction Crisis: Loss of Animal Populations and Species." *Journal of Cosmology* 8 (June).

"Center to Battle Future Food Crises." 2011. News of the Week. *Science* (October 14): 160–61.

Chakrabarty, Dipesh. 2009. "The Climate of History: Four Theses." *Critical Inquiry* 35:197–222.

Chesler, Ellen. 2010. "Women in the Center." In *A Pivotal Moment: Population, Justice, and the Environmental Challenge*, edited by L. Mazur, 311–20. Washington, DC: Island Press.

Chew, Sing. 2001. *World Ecological Degradation: Accumulation, Urbanization, and Deforestation, 3000BC–AD2000*. Walnut Creek, CA: AltaMira Press.

Clay, Jason. 2011. "Freeze the Footprint of Food." *Nature* 475 (July 21).

Coetzee, J. M. 2000. *Disgrace: A Novel*. Penguin Books.

Coetzee, J. M. 2001. *The Lives of Animals*. Princeton, NJ: Princeton University Press.

Coffin, Alisa. 2007. "From Road Kill to Road Ecology: A Review of the Ecological Effect of Roads." *Journal of Transport Geography* 15:396–406.

Cooper, Alan, et al. 2015. "Abrupt Warming Events Drove Late Pleistocene Holarctic Megafaunal Turnover." *Science* 349 (6248): 602–6.

Corlett, Richard. 2016. "Plant Diversity in a Changing World: Status, Trends, and Conservation Needs." *Plant Diversity* 38:10–16.

Cornwall, Warren. 2014. "The Missing Monarchs: Monsanto's Round-Up and Genetically Modified Crops Are Harming Everybody's Favorite Butterfly." *Slate*, January 29. http://www.slate.com/articles/health_and_science/ science/2014/01/monarch_butterfly_decline_monsanto_s_roundup_is _killing_milkweed.html.

Crist, Eileen. 2013a. "Ecocide and the Extinction of Animal Minds." In *Ignoring Nature No More: The Case for Compassionate Conservation*, edited by M. Bekoff, 45–61. Chicago: University of Chicago Press.

Crist, Eileen. 2013b. "On the Poverty of Our Nomenclature." *Environmental Humanities* 3:129–47.

Crist, Eileen. 2015. "I Walk in the World to Love It." In *Protecting the Wild: Parks and Wilderness, the Foundation for Conservation*, edited by George Wuerthner, Eileen Crist, and Tom Butler, 82–95. Washington, DC: Island Press.

Crist, Eileen, and H. Bruce Rinker, eds. 2009. *Gaia in Turmoil: Climate Change, Biodepletion, and Earth Ethics in an Age of Crisis*. Cambridge, MA: MIT Press.

Cronon, William. 1995. "The Trouble with Wilderness; or, Getting Back to the Wrong Nature." In *Uncommon Ground: Rethinking the Human Place in Nature*, ed. W. Cronon, 69–90. New York: W. W. Norton.

Crosby, Alfred. 2015. *Ecological Imperialism: The Biological Expansion of Europe, 900–1900*. 2nd ed. Cambridge: Cambridge University Press.

Crutzen, Paul, and Christian Schwägerl. 2011. "Living in the Anthropocene: Toward a New Global Ethos." *Yale Environment 360*, January 24.

Curry, Patrick. 2012. *Ecological Ethics: An Introduction*. Cambridge: Polity Press.

Daigle, Patrick. 2010. "A Summary of the Environmental Impacts of Roads, Management Responses, and Research Gaps: A Literature Review." *British Columbia Journal of Ecosystems and Management* 10 (3): 65–89.

Daily, Gretchen, and Paul Ehrlich. 1994. "Population, Sustainability, and Earth's Carrying Capacity." *Bioscience* 42 (10): 761–71.

Danson, Ted. 2011. *Oceana: Our Endangered Ocean and What We Can Do to Save It*. New York: Rodale.

Darwin, Charles. (1859) 1964. *Origin of Species*. Cambridge, MA: Harvard University Press.

Darwin, Charles (1871) 1981. *The Descent of Man, and Selection in Relation to Sex*. Princeton, NJ: Princeton University Press.

Dauvergne, Peter. 2008. *The Shadows of Consumption: Consequences for the Global Environment*. Cambridge, MA: MIT Press.

Dawkins, Marian Stamp, and Roland Bonney, eds. 2008. *The Future of Animal Farming: Renewing the Ancient Contract*. Blackwell Publishing.

Debord, Guy. 1995. *The Society of Spectacle*. New York: Zone Books.

deBuys, William. 2015. "The Politics of Extinction: An Introduction to the Most Beautiful Animal You'll Never See." *Huffington Post*, March 16, 2015. http://www.huffingtonpost.com/william-debuys/the-politics-of-extinction_b_6877156.html.

DellaSala, Dominique, et al. 2018. "The Anthropocene: How the Great Acceleration Is Transforming the Planet at Unprecedented Levels." *Encyclopedia of the Anthropocene*. https://doi.org/10.1016/B978-0-12-809665-9.09957-2.

Denevan, William. 1992. "The Pristine Myth: The Landscape of the Americas in 1492." *Annals of the Association of American Geographers* 82 (3): 369–85.

Derrida, Jacques. 1987. "The Ends of Man." In *After Philosophy: End or Transformation?*, edited by K. Baynes et al., 125–58. Cambridge, MA: MIT Press.

Derrida, Jacques. 2008. *The Animal That Therefore I Am*. New York: Fordham University Press.

Descola, Philippe. 2009. "Human Natures." *Social Anthropology* 17 (2): 145–57.

Diaz, Robert, and Rutger Rosenberg. 2008. "Spreading Dead Zones and Consequences for Marine Ecosystems." *Science* 321 (5891): 926–29.

Dinerstein, Eric. 2016. *The Return of the Unicorns: The Natural History and Conservation of the Greater-Horned Rhinoceros*. New York: Columbia University Press.

Dinerstein, Eric, et al. 2017. "An Ecoregion-Based Approach to Protecting Half the Terrestrial Realm." *Bioscience* 67 (6): 534–45.

Dirzo, Rodolfo, and Peter Raven. 2003. "Global State of Biodiversity and Loss." *Annual Review of Environmental Resources* 28:137–67.

Dirzo, Rodolfo, et al. 2014. "Defaunation in the Anthropocene." *Science* 345 (6195): 401–6.

do Sul, Juliana A. Ivar, and Monica Costa. 2014. "The Present and Future of Microplastic Pollution in the Marine Environment." *Environmental Pollution* 185:352–64.

Drengson, Alan. 2004. "The Wild Way." *Trumpeter* 20 (1): 46–65.

Dudgeon, David. 2010. "Requiem for a River: Extinctions, Climate Change, and the Last of the Yangtze." Editorial. *Aquatic Conservation: Marine and Freshwater Ecosystems* 20:127–31.

Dudgeon, David, et al. 2006. "Freshwater Biodiversity: Importance, Threats, Status and Conservation Challenges." *Biological Review* 81:163–82.

Dulac, John. 2013. "Global Land Transport Infrastructure Requirements: Estimating Road and Railway Infrastructure Capacity and Costs to 2050." An Information Paper. Paris: International Energy Agency.

Duncan, Emma. 2013. "All Creatures Great and Small." Special Report on Biodiversity. *Economist*, September 14, 3–16.

Durning, Alan. 1992. *How Much Is Enough? The Consumer Society and the Future of the Earth*. New York: W. W. Norton.

Eckersley, Robin. 2005. "Ecocentric Discourses: Problems and Future Prospects for Nature Advocacy." In *Debating the Earth: The Environmental Politics Reader*, edited by D. S. Dryzek and D. Schlosberg, 364–81. Oxford: Oxford University Press.

Ehrlich, Paul, and Anne Ehrlich. 2004. *One with Nineveh: Politics, Consumption, and the Human Future*. Washington, DC: Island Press.

Ehrlich, Paul, and Anne Ehrlich. 2006. "Enough Already." *New Scientist*, September 30, 47–50.

Ehrlich, Paul, and Anne Ehrlich. 2013. "Can a Collapse of Global Civilization Be Avoided?" *Proceedings of the Royal Society B* 280:20122845.

Ehrlich, Paul, and John Holdren. 1971. "The Impact of Population Growth" *Science* 171: 1212–17.

Ehrlich, Paul, and Robert Pringle. 2008. "Where Does Biodiversity Go from Here? A Grim Business-as-Usual Forecast and a Hopeful Portfolio of Partial Solutions." *PNAS* 105 (1): 11579–86.

Ehrlich, Paul, and John Harte. 2015. "Food Security Requires a New Revolution." *International Journal of Environmental Studies* 72 (6): 908–20.

Ellis, Erle. 2011a. "Anthropogenic Transformation of the Terrestrial Biosphere." *Philosophical Transactions of the Royal Society A* 369:1010–35.

Ellis, Erle. 2011b. "Neither Good Nor Bad." Editorial. *New York Times*, May 23. http://www.nytimes.com/roomfordebate/2011/05/19/the-age-of-anthropocene-should-we-worry/neither-good-nor-bad.

Ellis, Erle. 2012. "The Planet of No Return: Human Resilience on an Artificial Earth." Breakthrough Institute. http://thebreakthrough.org/index.php/journal/past-issues/issue-2/the-planet-of-no-return.

Ellis, Erle. 2013a. "The Long Anthropocene: Three Millennia of Humans Reshaping the Earth." http://thebreakthrough.org/index.php/programs/conservation-and-development/the-long-anthropocene.

Ellis, Erle. 2013b. "Overpopulation Is Not the Problem." *New York Times*, September 13.

Ellis, Erle, et al. 2013. "Used Planet: A Global History." *PNAS* 110 (20): 7978–85.

Engelman, R. 2008. *More: Population, Nature, and What Women Want*. Washington, DC: Island Press.

Engelman, Robert. 2012a. "Nine Population Strategies to Stop Short of 9 Billion." In *State of the World 2012: Moving toward Sustainable Prosperity*, 121–28. Washington, DC: Island Press.

Engelman, Robert. 2012b. "Trusting Women to End Population Growth." In *Life on the Brink: Environmentalists Confront Overpopulation*, edited by P. Cafaro and E. Crist, 223–39. Atlanta: University of Georgia Press.

Engelman, Robert. 2016. "Africa's Population Will Soar Dangerously unless Women Are More Empowered." *Scientific American*, February 1.

Erle, Sylvia. 2009. *The World Is Blue: How Our Fate and the Ocean's Is One*. Washington, DC: National Geographic Society.

Estes, James, et al. 2011. "Trophic Downgrading of Planet Earth." *Science* 333 (July 15): 301–6.

Estrada, Alejandro, et al. (2017). "Impending Extinction Crisis of the World's Primates: Why Primates Matter." *Science Advances* (Review) January 18. 3: e1600946.

Evans, Brock. 2015. "The Fight for Wilderness Preservation in the Pacific Northwest." In *Protecting the Wild: Parks and Wilderness, the Foundation for Conservation*, edited by G. Wuerthner, E. Crist, and T. Butler, 53–62. Washington, DC: Island Press.

Evernden, Neil. (1985) 1999. *The Natural Alien: Humankind and Environment*. 2nd ed. Toronto: Toronto University Press.

Ezeh, Alex, John Bongaarts, and Blessing Mberu. 2012. "Global Population Trends and Policy Options." *Lancet* 380:142–48.

Fabic, Madeleine, et al. 2015. "Meeting Demand for Family Planning within a Generation: The Post-2015 Agenda." *Lancet* 385:1928–31.

FAO (Food and Agriculture Organization). 2008. The State of the World's Fisheries and Aquaculture. https://www.scribd.com/doc/18420227/FAO-The-State-of-World-Fisheries-and-Aquaculture.

FAO (Food and Agriculture Organization). 2015. "Meeting the 2015 International Hunger Targets: Taking Stock of Uneven Progress." The State of Food Insecurity in the World. Rome. http://www.fao.org/3/a-i4646e.pdf.

Federoff, Nina, and Nancy Marie Brown. 2010. "Food for Thought." In *Technology and Values: Essential Readings*, edited by C. Hanks. Malden, MA: Wiley-Blackwell.

Feltman, Rachel. 2015. "Record Numbers of Starving Baby Seals Continue to Wash Ashore in California." *Washington Post*, March 16.

Fernandez, Carlos, and Andrea Nogués. 2010. "Preserving Patagonian Grasslands and Gauchos." *PERC Report* 28 (1). https://www.perc.org/articles/preserving-patagonian-grasslands-and-gauchos.

Foer, Jonathan Safran. 2009. *Eating Animals*. New York: Back Bay Books.

Foley, Jonathan, et al. 2005. "Global Consequences of Land Use." *Science* 309:570–74.

Foley, Jonathan, et al. 2011. "Solutions for a Cultivated Planet." *Nature* 478:337–42.

Forbes, Jack. (1979) 2008. *Columbus and Other Cannibals*. New York: Seven Stories Press.

Foreman, Dave. 2004. *Rewilding North America: A Vision for Conservation in the 21st Century*. Washington, DC: Island Press.

Francis (Pope). 2015. *Encyclical on Climate Change and Inequality: On Care of Our Common Home*. Brooklyn, NY: Melville House.

Freeman, Andrea. 2007. "Fast Food: Oppression through Poor Nutrition." *California Law Review* 95 (6): 2221–59.

Freese, Curt. 2015. "A New Era of Protected Areas for the Great Plains." In *Protecting the Wild: Parks and Wilderness, the Foundation for Conservation*, edited by George Wuerthner, Eileen Crist, and Tom Butler, 208–14. Washington, DC: Island Press.

Fukuoka, Masanobu. (1978) 2009. *The One-Straw Revolution*. New York: New York Review Books.

Fuller, Buckminster. 1982. *Synergetics: Explorations in the Geometry of Thinking*. New York: Macmillan.

Fulmer, Jeffrey. 2009. "What in the World Is Infrastructure?" *Infrastructure Investor*, July/August, 30–32.

Georgescu-Roegen, Nicholas. 1975. "Energy and Economic Myths." *Southern Economic Journal* 41 (3): 347–81.

Gerland, Patrick, et al. 2014. "World Population Stabilization Unlikely This Century." *Science* 346 (6206): 234–37.

Gibbs, H. K., et al. 2010. "Tropical Forests Were the Primary Sources of New Agricultural Land in the 1980s and 1990s." *PNAS* 107 (38): 16732–37.

Gjerde, Kristina, Duncan Currie, Kateryna Wowk, and Karen Sack. 2013. "Ocean in Peril: Reforming the Management of Global Ocean Living Resources in Areas beyond National Jurisdiction." *Marine Pollution Bulletin* 74 (2): 540–51.

Goldberg, Michelle. 2009. *The Means of Reproduction: Sex, Power, and the Future of the World*. New York: Penguin Press.

Goodall, Jane (with Phillip Berman). 1999. *Reason for Hope: A Spiritual Journey*. New York: Warner Books.

Grayson, Donald, and David Meltzer. 2003. "A Requiem for North American Overkill." *Journal for Archaeological Science* 30:585–93.

Griscom, Bronson, et al. 2017. "Natural Climate Solutions." *PNAS* (early edition). http://www.pnas.org/content/114/44/11645.

Guha, Ramachandra. 1989. "Radical American Environmentalism and Wilderness Preservation: A Third World Critique." *Environmental Ethics*, Spring, 71–83.

Habermas, Jürgen. 1984. *Theory of Communicative Action*. Vol. 1. Boston: Beacon Press.

Halweil, Brian. 2006. "Can Organic Farming Feed Us All?" *Worldwatch Magazine* 19 (3).

Halweil, Brian, and Danielle Nierenberg. 2008. "Meat and Seafood: The Global Diet's Most Costly Ingredients." In *State of the World 2008: Innovations for a Sustainable Economy*, chap. 5. Worldwatch Institute. New York: W. W. Norton.

Hamilton, Clive. 2014. *Earthmasters: The Dawn of the Age of Climate Engineering*. New Haven, CT: Yale University Press.

Hamilton, Clive, Christophe Bonneuil, and Francois Gemenne, eds. 2015. *The Anthropocene and the Global Environmental Crisis*. London and New York: Routledge.

Hance, Jeremy. 2008. "Marine 'Dead Zones' Double Every Decade." *Mongabay*, August 14. https://news.mongabay.com/2008/08/marine-dead-zones -double-every-decade/.

Hannibal, Mary Ellen. 2016. *Citizen Science: Searching for Heroes and Hope in an Age of Extinction*. New York: The Experiment.

Hansen, James, and Fred Gale. 2014. "China in the Next Decade: Rising Meat Demand and Growing Imports of Feed." http://www.ers.usda.gov/amber -waves/2014-april/china-in-the-next-decade-rising-meat-demand-and -growing-imports-of-feed.aspx#.V1LYEuSPyC4.

Haraway, Donna. 2008. *When Species Meet*. Minneapolis: University of Minnesota Press.

Haraway, Donna. 2016. *Staying with the Trouble: Making Kin in the Chthulucene*. Durham, NC: Duke University Press.

Haven, Kendall. 2006. *100 Greatest Science Inventions of All Time*. Westport, CT: Libraries Unlimited.

Hawken, Paul. 2002. "Commerce and Wilderness." *Wild Earth* 12 (3): 8–11.

Hawken, Paul. 2007. *Blessed Unrest*. London: Penguin Books.

Heidegger, Martin. (1947) 2013. "Letter on Humanism." In *From Modernism to Postmodernism: An Anthology*, edited by L. Cahoone, 274–308. Oxford: Blackwell.

Heidegger, Martin. 1975. *Poetry, Language, Thought*. New York: Perennial Library.

Heidegger, Martin. 1977a. "On the Question Concerning Technology." In *The Question Concerning Technology and Other Essays*, 3–35. New York: Harper & Row.

Heidegger, Martin. 1977b. "The Age of World Picture." In *The Question Concerning Technology and Other Essays*, 115–54. New York: Harper & Row.

Heidegger, Martin. 1977c. "The Turning." In *The Question Concerning Technology and Other Essays*, 36–49. New York: Harper & Row.

Heidegger, Martin. 1977d. "The Word of Nietzsche." In *The Question Concerning Technology and Other Essays*. New York: Harper & Row, 53–112.

Henning, Brian. 2016. "From the Anthropocene to the Ecozoic: Philosophy and Global Climate Change." *Midwest Studies in Philosophy* 40:284–95.

Hettinger, Ned. 2014. "Age of Man Environmentalism and Respect for an Independent Nature." Manuscript. http://hettingern.people.cofc.edu/Hettinger _Age_of_Man_Environmentalism.pdf.

Hill, Taylor. 2015. "5 Epic Migrations under Threat from Human Roadblocks." *TakePart*, March 31. http://www.takepart.com/article/2015/03/31/ migration-and-threats.

Hiss, Tony. 2014. "Can the World Really Set Aside Half of the Planet for Wildlife?" *Smithsonian*, September. http://www.smithsonianmag.com/science -nature/can-world-really set-aside-half-planet-wildlife-180952379/?no-ist.

Hoban, Russell. 1980. *Riddley Walker*. New York: Washington Square Press.

Hoekstra, Jonathan, Timothy Boucher, Taylor Ricketts, and Carter Roberts. 2005. "Confronting a Biome Crisis: Global Disparities of Habitat Loss and Protection." *Ecology Letters* 8:23–29.

Hopkins, Keith. 1983. "Murderous Games: Gladiator Conquests in Ancient Rome." *History Today* 66 (3).

Horkheimer, Max, and Theodor Adorno. 1972. *Dialectic of Enlightenment*. New York: Continuum.

House, Freeman. 1999. *Totem Salmon: Life Lessons from Another Species*. Boston: Beacon Press.

Huey, Raymond, Jonanthan Losos, and Craig Moritz. 2010. "Are Lizards Toast?" *Science* 328 (May 14): 832–33.

Huxley, Thomas. (1889) 2012. "Agnosticism." In *Collected Essays*, vol. 5, *Science and the Christian Tradition*. Cambridge: Cambridge University Press.

Imhoff, Daniel, ed. 2010. *The CAFO Reader: The Tragedy of Industrial Animal Factories*. Berkeley, CA: Watershed Media.

Invisible Committee. 2009. *The Coming Insurrection*. Los Angeles: Semiotext(e).

Jackson, Jeremy. 2005. "When Ecological Pyramids Were Upside Down." In *Whales, Whaling, and Ocean Ecosystems*, edited by J. A. Estes, 27–37. Berkeley: University of California Press.

Jackson, Jeremy. 2008. "Ecological Extinction and Evolution in the Brave New Ocean." *PNAS* 105 (1): 11458–65.

Jackson, Jeremy, and Karen Alexander. 2011. "Introduction: The Importance of Shifting Baselines." In *Shifting Baselines: The Past and Future of Ocean Fisheries*, edited by J. Jackson, K. Alexander, and E. Sala, 1–11. Washington, DC: Island Press.

Jackson, Jeremy, Karen Alexander, and Enrik Sala, eds. 2011. *Shifting Baselines: The Past and Future of Ocean Fisheries*. Washington, DC: Island Press.

Jackson, Wes. 2011. *Nature as Measure*. Berkeley, CA: Counterpoint.

Jamieson, Alan et al. 2017. "Bioaccumulation of Persistent Organic Pollutants in the Deepest Ocean Fauna." *Nature Ecology & Evolution* 1, article 0051.

Jaworowski, Matt. 2015. "New Analysis: 1 in 6 Could Face Extinction Due to Global Warming." http://wiat.com/2015/05/05/new-analysis-1-in-6-species-could-face-extinction-due-to-global-warming/.

Jelinski, D. E. 2005. "There Is No Mother Nature—There Is No Balance of Nature: Culture, Ecology and Conservation." *Human Ecology* 33 (2): 271–88.

Jenkins, Martin. 2003. "Prospects for Biodiversity." *Science* 302: 1175–77.

Jensen, Derrick. 2002. *Listening to the Land: Conversations about Nature, Culture, and Eros*. White River Junction, VT: Chelsea Green.

Jensen, Derrick. 2006. *Endgame*. Vols. 1 and 2. New York: Seven Stories.

Jensen, Derrick. 2013. "Age of the Sociopath." *Earth Island Journal*, Spring, 41.

Jensen, Derrick. 2016. *The Myth of Human Supremacy*. New York: Seven Stories.

Johns, David. 2002. "Wilderness and Energy: The Battle against Domination." *Wild Earth*, Fall, 12–13.

Johns, David. 2016. "Rewilding." *Reference Module in Earth Systems and Environmental Sciences*, February 8, 1–11. DOI: 10.1016/B978-0-12-409548 -9.09202-2.

Jonas, Hans. 1974. *Philosophical Essays*. Englewood Cliffs, NJ: Prentice-Hall.

Jones, Mark. 2010. "Is Africa's Wildlife Being Eaten to Extinction?" *BBC News*, August 3. http://news.bbc.co.uk/2/hi/science/nature/8877062.stm.

Jones, Nicola. 2011. "Human Influence Comes of Age." *Nature* 473:133.

Kaidbey, Mona, and Robert Engelman. 2017. "Our Bodies, Our Future: Expanding Comprehensive Sexuality Education." In *EarthEd: Rethinking Education on a Changing Planet*, chap. 15. State of the World 2017. Worldwatch Institute. Washington, DC: Island Press.

Kareiva, Peter, Robert Lalasz, and Michelle Marvier. 2011. "Conservation in the Anthropocene: Beyond Solitude and Fragility." *Breakthrough Journal*, Fall, 29–37.

Kareiva, Peter, and Michelle Marvier. 2012. "What Is Conservation Science?" *Bioscience* 62 (11): 962–69.

Keim, Brandon. 2014. "Earth Is Not a Garden." *Aeon*, September 18. https://aeon .co/essays/giving-up-on-wilderness-means-a-barren-future-for-the-earth.

Keith, Lierre. 2011. "Other Plans." In *Deep Green Resistance: Strategy to Save the Planet*, edited by Aric McBay, Keith Lierre, and Derrick Jensen, 193–237. New York: Seven Stories.

Ketcham, Christopher. 2017a. "Inside the Effort to Kill Protections for Endangered Animals." *National Geographic*, May 19. http://news .nationalgeographic.com/2017/05/endangered_speciesact/.

Ketcham, Christopher. 2017b. "The Fallacy of Endless Economic Growth." *Pacific Standard*, May 16. https://psmag.com/magazine/fallacy-of-endless -growth.

Kharas, H. 2017. *The Unprecedented Expansion of the Global Middle Class: An Update*. Brookings Global Economy and Development. Working Paper 100.

Kidd, Elliott. 2012. "Beast-Hunts in Roman Amphitheaters: The Impact of the Venationes on Animal Populations in the Ancient Roman World." *Eagle Feather*, http://eaglefeather.honors.unt.edu/2012/article/32#.V77SI6KPzE8.

Kidner, David. 2014. "Why 'Anthropocentrism' Is Not Anthropocentric." *Dialectical Anthropology* 38 (4): 465–80.

Kimmerer, Robin. 2003. *Gathering Moss: A Natural and Cultural History of Mosses*. Corvallis: Oregon State University Press.

Kimmerer, Robin. 2010. "The Giveaway." In *Moral Ground*, edited by Kathleen Dean Moore and Michael Nelson, 141–45. San Antonio, TX: Trinity University Press.

Kingsnorth, Paul. 2013. "Dark Ecology." *Orion*, January/February.

Kirschenmann, Fred. 2008. "Food as Relationship." *Journal of Hunger and Environmental Nutrition* 3 (2/3): 106–21.

Kirschenmann, Fred. 2017. "From Soil to Sustainability." https://slowmoney .org/blog/from-soil-to-sustainability.

Klare, Michael. 2012. *The Race for What's Left: The Global Scramble for the World's Last Resources*. New York: Picador.

Kloppenburg, Jack, John Hendrickson, and G. W. Stevenson. 1996. "Coming into the Foodshed." *Agriculture and Human Values* 13 (3): 33–42.

Koch, Paul, and Anthony Barnosky. 2006. "Late Quaternary Extinctions: State of the Debate." *Annual Review of Ecology, Evolution, and Systematics* 37:215–50.

Kohák, Erazim. 1987. *The Ember and the Stars: A Philosophical Inquiry into the Moral Sense of Nature*. Chicago: University of Chicago Press.

Kolbert, Elizabeth. 2011. "Enter the Anthropocene—Age of Man," *National Geographic*, March.

Kolbert, Elizabeth. 2014. *The Sixth Extinction*. New York: Henry Holt and Co.

Kopnina, Helen. 2012. "Re-examining Culture/Conservation Conflict: The View of Anthropology of Conservation through the Lens of Environmental Ethics." *Journal of Integrative Environmental Sciences* 9 (1): 9–25.

Koslow, Tony. 2007. *The Silent Deep: The Discovery, Ecology and Conservation of the Deep Sea*. Chicago: University of Chicago Press.

Kotchen, Matthew, and Oran Young. 2007. "Meeting the Challenges of the Anthropocene: Towards a Science of Coupled Human-Biophysical Systems." *Global Environmental Change* 17:149–51.

Kottasova, Ivana. 2016. "More Plastic Than Fish in Oceans by 2050." http://money.cnn.com/2016/01/19/news/economy/davos-plastic-ocean-fish/index.html?iid=hp-stack-dom.

Krall, Lisi. 2007. "Between Wilderness and the Middle Landscape: A Rocky Road." In *US Forest Service Proceedings* RMRS-P-49, 134–40.

Krall, Lisi. 2015. "The Economic Evolution of Dominion." https://www.youtube.com/watch?v=r8VNSrl6yck.

Kraus, Bernie. 2012. *The Great Animal Orchestra: Finding the Origins of Music in the World's Wild Places*. London: Profile Books.

Kroll, Gary. 2015. "An Environmental History of Roadkill: Road Ecology and the Making of the Permeable Highway." *Environmental History* 20:4–28.

Kurlansky, Mark. 2011. *World without Fish*. New York: Workman Publishing Company.

LaDuke, Winona. 1997. "Voices from White Earth: Gaa-waabaabiganikaag." In *People, Land, and Community*, edited by H. Hannum, 22–37. New Haven, CT: Yale University Press.

Laurance, William. 2015. "Roads to Ruin." Opinion. *New York Times*, April 12.

Laurance, William, et al. 2015. "Estimating the Environmental Costs of Africa's Massive 'Development Corridors.'" *Current Biology* 25:3202–08.

Law, Kara Lavender, and Richard Thompson. 2014. "Microplastics in the Seas." *Science* 345 (6193): 144–45.

Lee, Tien Ming, et al. 2014. "The Harvest of Wildlife for Bushmeat and Traditional Medicine in East, South and Southeast Asia." Occasional Paper 115. Center for International Forestry Research.

Lees, Alexander, and Stuart Pimm. 2015. "Species, Extinct before We Know Them?" *Current Biology* 25 (5): 177–80.

Lenton, Timothy. 2004. "Clarifying Gaia: Regulation with or without Natural Selection." In *Scientists Debate Gaia: The Next Century,* edited by S. Schneider, J. Miller, E. Crist, and P. Boston, 15–25. Cambridge, MA: MIT Press.

Lenzen, M., et al. 2012. "International Trade Drives Biodiversity Threats in Developing Nations." *Nature* 486:109–12.

Leopold, Aldo. (1949) 1968. *A Sand County Almanac and Sketches Here and There.* Oxford: Oxford University Press.

Lester, Lori, et al. 2016. "Use of a Florida Gulf Coast Barrier Island by Spring Trans-Gulf Migrants and the Projected Effects of Sea Level Rise on Habitat Availability." *PLOS One,* March 2. DOI: 10.1371/journal.pone.0148975.

Locke, Harvey. 2013. "Nature Need Half: A Necessary and Hopeful New Agenda for Protected Areas." *Parks* 19 (2).

Louv, Richard. 2008. *Last Child in the Woods: Saving Our Children from Nature Deficit Disorder.* Chapel Hill, NC: Algonquin Books.

Lovejoy, Arthur. (1936) 1976. *The Great Chain of Being: A Study of the History of an Idea.* Cambridge, MA: Harvard University Press.

Lovelock, James. 1988. *The Ages of Gaia.* New York: W. W. Norton and Company.

Lovgren, Stefan. 2006. "China's Rare River Dolphin Now Extinct Experts Announce." *National Geographic News,* December 14. http://news.nationalgeographic.com/news/2006/12/061214-dolphin-extinct.html.

Lulka, David. 2004. "Stabilizing the Herd: Fixing the Identity of Nonhumans." *Environment and Planning D; Society and Space* 22:439–63.

Lynas, Mark. 2011. *The God Species: How the Planet Can Survive the Age of Humans.* London: Fourth Estate.

Machovina, Brian, Kenneth Feeley, and William Ripple. 2015. "Biodiversity Conservation: The Key Is Reducing Meat Consumption." *Science of the Total Environment* 536:419–31.

MacKay, Paula. 2001. "Farming with the Wild: Reconnecting Food Systems with Ecosystems." *Wild Earth* 11 (2): 52–55.

Mackey, Brendan, et al. 2015. "Policy Options for the World's Primary Forests in Multilateral Environmental Agreements." *Conservation Letters* 8 (2): 139–47.

MacKinnon, Michael. 2006. "Supplying Exotic Animals for the Roman Amphitheatre Games: New Reconstructions Combining Archeological, Ancient Textual, Historical and Ethnographic Data." *Mouseion* 6:1–25.

Madhuri, S., et al. 2014. "Fish Cancer Developed by Environmental Pollutants." *International Research Journal of Pharmacy* 3 (10): 17–19.

Magdoff, Fred, and John Bellamy Foster. 2011. *What Every Environmentalist Needs to Know about Capitalism: A Citizen's Guide to Capitalism and Environment.* New York: Monthly Review Press.

Mander, Damien. 2013. "Modern Warrior." TED talk. http://www.iapf.org/en/about/blog/entry/modern-warrior-damien-mander-at-tedxsydney.

Mander, Jerry. 1992. *In the Absence of the Sacred: The Failure of Technology and the Survival of the Indian Nations.* San Francisco: Sierra Club Books.

Mander, Jerry, ed. 2007. *Manifesto on Global Economic Transitions.* Project of the International Forum on Globalization. Institute for Policy Studies.

Mander, Jerry, and Victoria Tauli-Corpuz, eds. 2006. *Paradigm Wars: Indigenous People's Resistance to Globalization.* San Francisco: Sierra Club Books.

Mann, Charles. 2002. "1491." *Atlantic,* March. http://www.theatlantic.com/magazine/archive/2002/03/1491/302445/.

Mann, Charles. (2005) 2011. *1491: New Revelations of the Americas before Columbus.* 2nd ed. New York: Vintage Books.

Manning, Richard. 2004. *Against the Grain: How Agriculture Hijacked Civilization.* New York: North Point Press.

Manning, Richard. 2013. "Bakken Business: The Price of North Dakota's Fracking Boom." *Harpers,* March.

Manning, Richard. 2016. "The Trouble with Iowa: Corn, Corruption, and the Presidential Caucuses." *Harpers,* February.

Marcuse, Herbert. 1969. *An Essay on Liberation.* Boston: Beacon Press.

Margulis, Lynn. 1993. "Gaia and the Colonization of Mars." *GSA Today* 3 (11): 277–91.

Margulis, Lynn. 1998. *Symbiotic Planet: A New View of Evolution.* New York: Basic Books.

Marris, Emma. 2009. "Ecology: Ragamuffin Earth." *Nature* 460:450–53.

Marris, Emma. 2013. "The New Wild." Lecture presented by the University of Montana Wilderness Institute. https://www.youtube.com/watch?v=HvWlQELBqN4.

Marshall, Robert. (1930) 1998. "The Problem of the Wilderness." In *The Great New Wilderness Debate,* edited by J. Baird Callicott and Michael Nelson, 85–96. Athens: University of Georgia Press.

Martin, Paul S. 1967. "Prehistoric Overkill." In *Pleistocene Extinctions: The Search for a Cause,* edited by Paul S. Martin and H. E. Wright Jr., 75–120. New Haven, CT: Yale University Press.

Marx, Leo. 1996. "The Domination of Nature and the Redefinition of Progress." In *Progress: Fact or Illusion?,* edited by L. Marx and B. Mazlish, 201–18. Ann Arbor: University of Michigan Press.

Mathews, Freya. 2011. "Planet Beehive." *Australian Humanities Review* 50. http://www.australianhumanitiesreview.org/archive/Issue-May-2011/mathews.html.

Maverick, Tim. 2014. "China's Hunger for Soy to Exceed Global Supply." *Wall Street Journal,* November 11. http://www.wallstreetdaily.com/2014/11/11/china-soybean-futures/.

Maxwell, Sean et al. 2016. "The Ravages of Guns, Nets and Bulldozers." *Nature* 536:143–45.

Mazoyer, Marcel, and Laurence Roudart. 2006. *A History of World Agriculture: From the Neolithic to the Current Crisis*. New York: Monthly Review Press.

McCauley, Douglas, et al. 2015. "Marine Defaunation: Animal Loss in the Global Ocean." *Science* 347 (6219): 247.

McKee, Jeffrey, and Erica Chambers. 2011. "Behavioral Mediators of the Human Population Effect on Global Biodiversity Losses." In *Human Population: Its Influences on Biological Diversity*, edited by R. P. Cincotta and L. J. Gorenflo, 47–59. Ecological Studies 214. Berlin: Springer-Verlag.

McKee, Jeffrey, Erica Chambers, and Julie Guseman. 2013. "Human Population Density and Growth Validated as Extinction Threats to Mammal and Bird Species." *Human Ecology*. DOI: 10.1007/s10745-013-9586-8.

McKibben, Bill. 1998. *Maybe One: A Case for Smaller Families*. New York: Plume.

McKibben, Bill. 2010. "The Only Way to Have a Cow." *Orion*. https://orionmagazine.org/article/the-only way-to-have-a-cow/.

McKibben, Bill. 2011. *Eaarth: Making a Life on a Tough New Planet*. New York: St. Martin's Griffin.

McKibben, Bill. 2016. "Recalculating the Climate Math." *New Republic*, September 22. https://newrepublic.com/article/136987/recalculating-climate-math.

McLuhan, T. C., ed. 1971. *Touch the Earth: A Self-Portrait of Indian Existence*. New York: Promontory Press.

Merchant, Carolyn. 1980. *The Death of Nature: Women, Ecology and the Scientific Revolution*. New York: HarperCollins.

Metzner, Ralph. 1993. "The Split between Spirit and Nature in European Consciousness." *Trumpeter* 10 (1).

Millennium Ecosystem Assessment. 2005. http://millenniumassessment.org/en/index.html.

Miller, James. 2005. "Biodiversity Conservation and the Extinction of Experience." *Trends in Ecology and Evolution* 20 (8): 430–34.

Miller, Jeremy. 2017. "Bounty Hunters: A Clandestine War on Wolves." *Harpers*, January, 69–75.

Mills, Stephanie. 1997. "Making Amends to the Myriad Creatures." In *People, Land, and Community*, edited by H. Hannum, 275–88. New Haven, CT: Yale University Press.

Mills, Stephanie. 2010. "Peak Nature?" *The Post Carbon Reader Series: Biodiversity*. https://www.scribd.com/document/63608422/BIODIVERSITY-Peak-Nature-by-Stephanie-Mills.

Monastersky, Richard. 2014. "Life—A Status Report." *Nature* 516:159–61.

Monbiot, George. 2013. *Feral: Searching for Enchantment on the Frontiers of Rewilding*. London: Allen Lane Books.

Monbiot, George. 2015. "There's a Population Crisis All Right. But Probably Not the One You Think." *Guardian*, November 19. https://www.theguardian.com/commentisfree/2015/nov/19/population-crisis-farm-animals-laying-waste-to-planet.

Monson, Shaun. 2010. *Earthlings*. DVD. Nation Earth.

Moore, Charles. 2009. "Sea of Plastic." http://www.ted.com/talks/capt_charles
_moore_on_the_seas_of_plastic?language=en.

Moore, Jason, ed. 2016a. *Anthropocene or Capitalocene? Nature, History, and the
Crisis of Capitalism*. Oakland, CA: PM Press.

Moore, Jason. 2016b. "The Rise of Cheap Nature." In *Anthropocene or Capitalocene? Nature, History, and the Crisis of Capitalism*, edited by J. Moore, 78–
115. Oakland, CA: PM Press.

Mora, Camilo, et al. 2011. "How Many Species Are There on Earth and in the
Ocean?" http://journals.plos.org/plosbiology/article?id=10.1371/journal
.pbio.1001127.

Morelle, Rebecca. 2016. "PCB Chemical Threat to Europe's Killer Whales and
Dolphins." http://www.bbc.com/news/science-environment-35302957.

Morello, Lauren. 2010. "Oceans Turn More Acidic Than Last 800,000 Years."
Scientific American, February 22. https://www.scientificamerican.com/
article/acidic-oceans/.

Mowat, Farley. (1984) 2004. *Sea of Slaughter*. Mechanicsburg, PA: Stackpole
Books.

Mumford, Lewis. 1964. "Authoritarian and Democratic Technics." *Technology
and Culture* 5 (1): 1–8.

Mumford, Lewis. 1967. *The Myth of the Machine: Technics and Human Development*. San Diego, CA: Harcourt, Brace & World Inc.

Myers, Norman, and Andrew Knoll. 2001. "The Biotic Crisis and the Future of
Evolution." *PNAS* 98 (10): 5389–92.

Myers, Ransom, and Boris Worm. 2005. "Extinction, Survival or Recovery of
Large Predatory Fishes." *Philosophical Transactions of the Royal Society B*
360: 13–20.

Naess, Arne. 1995. "Equality, Sameness, and Rights." In *Deep Ecology for the
21st Century*, edited by G. Sessions, 222–24. Boston: Shambhala.

Nash, Roderick. 1989. *The Rights of Nature*. Madison: University of Wisconsin
Press.

Nash, Roderick. 2012. "Island Civilization: A Vision of Human Inhabitance in
the Fourth Millennium." In *Life on the Brink: Environmentalists Confront
Overpopulation*, edited by P. Cafaro and E. Crist, 301–12. Atlanta: University of Georgia Press.

National Research Council. 1970. *Land Use and Wildlife Resources*. http://www
.ncbi.nlm.nih.gov/books/NBK208755/.

Nibert, David. 2012. "The Fire Next Time: The Coming Cost of Capitalism,
Animal Oppression, and Environmental Ruin." *Journal of Human Rights
and the Environment* 3 (1): 141–58.

Nibert, David. 2013. *Animal Oppression and Human Violence: Domesecration,
Capitalism, and Global Conflict*. New York: Columbia University Press.

Nietzsche, Friedrich. (1954) 1968. *The Portable Nietzsche*. New York: Viking
Press.

Nietzsche, Friedrich. 1974. *The Gay Science.* New York: Vintage Books.

Nimmo, Richie. 2011. "The Making of the Human: Anthropocentrism in Modern Social Thought." In *Anthropocentrism: Humans, Animals, Environment,* edited by Rob Boddice. Leiden: Brill.

Noske, Barbara. 1997. "Speciesism, Anthropocentrism, and Non-Western Cultures." *Anthrozoös* 10 (4): 183–90.

Noss, Reed. 1998. "The Ecological Effects of Roads." http://www.eco-action .org/dt/roads.html.

Noss, Reed. 2001. "Beyond Kyoto: Forest Management in a Time of Rapid Climate Change." *Conservation Biology* 15 (3): 578–90.

Noss, Reed. 2013. "Wilderness, Wildness, and Biodiversity: We Need All Three." Lecture presented by the University of Montana Wilderness Institute. https://www.youtube.com/watch?v=UcCUX_w5uhw .

Noss, Reed, et al. 2011. "Bolder Thinking for Conservation." *Conservation Biology* 26 (1): 1–4.

Noss, Reed, and Allan Cooperrider. 1994. *Saving Nature's Legacy.* Washington, DC: Island Press.

Nye, David. 2003. "Technology, Nature, and American Origin Stories." *Environmental History* 8 (1): 8–24.

Orr, David. 2004. *Earth in Mind: On Education, Environment, and the Human Prospect.* 2nd ed. Washington, DC: Island Press.

Osborn, Ronald. 2010. "Seyla Benhabib, Wendell Berry, and the Question of Migrant and Refugee Rights." *Humanitas* 23 (1/2): 118–38.

Ottaway, Richard, et al. 2007. *Return of the Population Growth Factor: Its Impact upon the Millennium Development Goals.* Report of Hearings by the All Party Parliamentary Group on Population, Development and Reproductive Health. London: House of Commons. http://www.appgpopdevrh.org.uk/ Publications/Population%20Hearings/APPG%20Report%20-%20Return %20of%20the%20Population%20Factor.pdf.

Ouspensky, P. D. (1950) 1977. *In Search of the Miraculous.* London: Routledge and Kegan Paul.

Owen, James. 2010. "New Earth Epoch Has Begun, Scientists Say," *National Geographic News,* April 6.

Palmer, Brian. 2015. "There Goes the Neighborhood." https://www.nrdc.org/ onearth/there-goes-neighborhood.

Patel, Raj. 2009. "What Does Food Sovereignty Look Like?" *Journal of Peasant Studies* 36 (3): 663–706.

Patel, Raj. 2011. *The Value of Nothing: How to Reshape Market Society and Redefine Democracy.* London: Portobello.

Patel, Raj. 2012. *Stuffed and Starved: The Hidden Battle for the World Food System.* Brooklyn, NY: Melville House.

Pauly, Daniel, and Dirk Zeller. 2016. "Catch Reconstructions Reveal That Global Marine Fisheries Catches Are Higher Than Reported and Declining." *Nature Communications* 7 (10244). DOI: 10.1038/ncomms10244.

Payne, Jonathan, et al. 2016. "Ecological Selectivity of the Emerging Mass Extinction in the Oceans." *Science*, September, 14. DOI: 10.1126/science. aaf2416.

Pearce, Fred. 2011. "Agribusiness Boom Threatens Key African Wildlife Migration." *Yale Environment 360*, March 7.

Pearce, Fred. 2013. "True Nature: Revising Ideas on What Is Pristine and Wild." *Yale Environment 360*, May 13.

Petersen, Dale. 2015. *Where Have All the Animals Gone? My Travels with Karl Ammann*. Peterborough, NH: Bauhan Publishing.

Petrini, Carlo. 2001. *Slow Food: The Call for Taste*. New York: Columbia University Press.

Pflanz, M. 2014. "The Ivory Police." *Christian Science Monitor*, March 2, 26.

Pimentel, David, et al. 2010. "Will Limited Land, Water, and Energy Control Human Population Numbers in the Future?" *Human Ecology* 38 (5): 599–611.

Pimm, Stuart. 2013. "Why Saving Species?" *BES Bulletin* 44 (4): 56–57.

Pimm, Stuart, et al. 2014. "The Biodiversity of Species and Their Rates of Extinction, Distribution, and Protection." *Science* 344 (6187). DOI: 10.1126/ science.1246752.

Plato. 1961. "Timaeus." In *The Collected Dialogues of Plato*, edited by E. Hamilton and H. Cairns, 1151–211. Princeton, NJ: Princeton University Press.

Platt, Jonathan. 2015. "African Lion Populations Drop 42 Percent in Past 21 Years." https://blogs.scientificamerican.com/extinction-countdown/ african-lion-populations-drop-42-percent-in-past-21-years/.

Plotica, Luke Philip. 2016. "Thoreau and the Politics of Ordinary Actions." *Political Theory* 44 (4): 470–95.

Plumwood, Val. 2008. "Shadow Places and the Politics of Dwelling." *Australian Humanities Review* 44. http://australianhumanitiesreview.org/archive/Issue -March-2008/plumwood.html.

Polidoro, Beth, et al. 2010. "The Loss of Species: Mangrove Extinction Risk and Geographic Areas of Global Concern." *PLOS One* 5 (4): 1–10.

Pollan, Michael. 2014. Foreword to *Grass, Soil, Hope: A Journey through Carbon Country*, by Courtney White. White River Junction, VT: Chelsea Green.

Ponting, Clive. (1991) 2007. *A New Green History of the World: The Environment and the Collapse of Great Civilizations*. London: Penguin Books.

Potts, Malcolm. 2009. "Where Next?" *Philosophical Transactions of the Royal Society B* 364:3115–24.

Pounds, Alan, et al. 2006. "Widespread Amphibian Extinctions from Epidemic Disease Driven by Global Warming." *Nature* 439 (January 12): 161–67.

Pretty, Jules. 2007. *The Earth Only Endures: On Reconnecting with Nature and Our Place in It*. London: Earthscan.

Punke, Michael. 2007. *Last Stand: George Bird Grinnell, the Battle to Save the Buffalo, and the Birth of the New West*. New York: Harper Collins.

Quammen, David. 2010. "Great Migrations." *National Geographic*, November. http://ngm.nationalgeographic.com/2010/11/great-migrations/quammen -text.

Ravallion, Martin. 2009. "The Developing World's Bulging but Vulnerable 'Middle Class.'" World Bank Development Research Group. Policy Research Working Paper 4816.

Redman, Charles. 1999. *Human Impact on Ancient Environments.* Tucson: University of Arizona Press.

Rees, William. 2014. "Avoiding Collapse: An Agenda for Sustainable Degrowth and Relocalizing the Economy." Canadian Centre for Policy Alternatives, British Columbia Office.

Reganold, John, and Jonathan Wachter. 2016. "Organic Agriculture in the 21st Century." *Nature Plants* 2:1–8.

Reid, Julian. 2012. "The Neoliberal Subject: Resilience and the Art of Living Dangerously." *Revista Pléyade* 10:143–65.

Revkin, Andrew. 2011. "Confronting the 'Anthropocene.'" *New York Times*, May 11.

Rinker, Bruce H. 2011. "Forest Systems and Gaia Theory." In *Gaia in Turmoil: Climate Change, Biodepletion, and Earth Ethics in an Age of Crisis*, ed. E. Crist and H. B. Rinker, 85–104. Cambridge: MIT Press.

Ripple, William, et al. 2014. "Status and Ecological Effects of the World's Largest Carnivores." *Science* 343 (6167): 124484. DOI: 10.1126/science.1241484.

Ripple, William, et al. 2015. "Collapse of the World's Largest Herbivores." Review. *Science Advances*, May 1.

Ripple, William, et al. 2016. "Bushmeat Hunting and Extinction Risk to the World's Mammals." www.pnas.org/cgi/doi/10.1073/pnas.1702078114.

Roberts, Callum. 2007. *The Unnatural History of the Sea.* Washington, DC: Island Press.

Roberts, Callum. 2012. *The Ocean of Life: The Fate of Man and the Sea.* New York: Penguin Books.

Rockström, Johan, et al. 2009. "A Safe Operating Space for Humanity." *Nature* 461, no. 24: 472–75.

Rodman, John. 1980. "Paradigm Change in Political Science: An Ecological Perspective." *American Behavioral Scientist* 24 (1): 49–78.

Rodman, John. 1987. "The Liberation of Nature?" *Inquiry* 20:83–145.

Rolston, Holmes. 1985. "The Exuberance of Life." *BioScience* 35 (11): 718–26.

Rolston, Holmes. 2015. "After Preservation? Dynamic Nature in the Anthropocene." In *After Preservation: Saving American Nature in the Age of Humans*, edited by B. Minteer and S. J. Pyne, 32–40. Chicago: University of Chicago Press.

Root, Terry, and Stephen Schneider. 2006. "Conservation and Climate Change: The Challenges Ahead." *Conservation Biology* 20 (3): 706–8.

Rorty, Richard. (1991) 1996. "Solidarity or Objectivity?" In *From Modernism to Postmodernism: An Anthology*, edited by L. Cahoone, 573–88. Oxford: Blackwell.

Rose, Deborah Bird. 2013. *Wild Dog Dreaming: Love and Extinction*. Charlottesville: University of Virginia Press.

Rose, Deborah Bird. 2014. "Arts of Flow: Poetics of 'Fit' in Aboriginal Australia." *Dialectical Anthropology* 38 (4): 431–45.

Ryerson, Bill. 2015. Introduction to *Overdevelopment, Overpopulation, Overshoot*, edited by T. Butler. Goff Books.

Safina, Carl. 2011. "A Shoreline Remembrance." In *Shifting Baselines: The Past and Future of Ocean Fisheries*, edited by Jeremy Jackson, Karen Alexander, and Enrik Sala, 13–19. Washington, DC: Island Press.

Sala, Enric, and Jeremy Jackson. 2011. "Lessons from Coral Reefs." In *Shifting Baselines: The Past and Future of Ocean Fisheries*, edited by Jeremy Jackson, Karen Alexander, and Enrik Sala, 193–203. Washington, DC: Island Press.

Sale, Kirkpatrick. 1990. *The Conquest of Paradise: Christopher Columbus and the Columbian Legacy*. New York: Alfred A. Knopf.

Sale, Kirkpatrick. 1997. "Mother of All: An Introduction to Bioregionalism." In *People, Land, and Community*, edited by H. Hannum, 216–35. New Haven, CT: Yale University Press.

Sale, Peter. 2011. *Our Dying Planet: An Ecologist's View of the Crisis We Face*. Berkeley: University of California Press.

Sanderson, Eric, et al. 2002. "The Human Footprint and the Last of the Wild." *BioScience* 52 (10): 891–904.

Sandom, Christopher, et al. 2014. "Global Late Quaternary Megafaunal Extinctions Linked to Humans, Not Climate Change." *Proceedings of the Royal Society B* 281:20133254.

Sax, Boria. 2011. "What Is This Quintessence of Dust? The Concept of the 'Human' and Its Origins." In *Anthropocentrism: Humans, Animals, Environment*, edited by Rob R. Boddice, 21–36. Leiden: Brill.

Schlosser, Eric. 2011. "Still a Fast Food Nation." http://www.thedailybeast.com/still-a-fast-food-nation-eric-schlosser-reflects-on-10-years-later.

Schor, Juliet. 2010. "Tackling Turbo Consumption." *Culture Studies* 22 (5): 588–98.

Schroll, Mark, and Heather Walker. 2011. "Diagnosing the Human Superiority Complex: Providing Evidence the Eco-Crisis in Born of Conscious Agency." *Anthropology of Consciousness* 22 (1): 39–48.

Schutz, Alfred. 1999. *On Phenomenology and Social Relations*. Edited and introduced by Helmut Wagner. Chicago: University of Chicago Press.

Scully, Matthew. 2003. *Dominion: The Power of Man, the Suffering of Animals, and the Call to Mercy*. New York: St. Martin's Press.

Seddon, Philip, et al. 2014. "Reversing Defaunation: Restoring Species in a Changing World." *Science* 345:406–12.

Sedgh, Gilda, et al. 2014. "Intended and Unintended Pregnancies Worldwide in 2012 and Recent Trends." *Studies in Family Planning* 45 (3): 301–14.

Sessions, George, ed. 1995. *Deep Ecology for the 21st Century*. Boston: Shambhala.

Shabekoff, Philip. 2003. *A Fierce Green Fire: The American Environmental Movement*. Washington, DC: Island Press.

Shellenberger, Michael, and Ted Nordhaus. 2011. *Love Your Monsters: Postenvironmentalism and the Anthropocene*. e-book. Oakland, CA: Breakthrough Institute.

Shukman, David. 2016. "Rhino Poaching: Another Year, Another Grim Record." http://www.bbc.com/news/science-environment-35769413.

Simard, Suzanne. 2016. "How Trees Talk to Each Other." TED Talk. https://www.ted.com/talks/suzanne_simard_how_trees_talk_to_each_other.

Slotkin, Richard. 1973. *Regeneration through Violence: The Mythology of the American Frontier, 1600–1860*. Middletown, CT: Wesleyan University Press.

Smail, Kenneth J. 1997. "Population Growth Seems to Affect Everything but Is Seldom Held Responsible for Anything." *Politics and the Life Sciences* 16 (2): 231–36.

Smeeding, Timothy. 2014. "Adjusting to the Fertility Bust." *Science* 346 (6206): 163–64.

Smil, Vaclav. 1997. "Global Population and the Nitrogen Cycle." *Scientific American*, July, 76–81.

Smil, Vaclav. 2011. "Harvesting the Biosphere: The Human Impact." *Population and Development Review* 37 (4): 613–36.

Smith, Gerald. 2002–3. "Distortion of World Fisheries by Capital." *Wild Earth* 12 (4): 16–17.

Smith, Martin, et al. 2010. "Sustainability and Global Seafood." *Science* 327:784–86.

Snyder, Gary. 1990. *The Practice of the Wild*. New York: North Point Press.

Sodhi, Navjot, et al. 2004. "Southeast Asian Biodiversity: An Impending Disaster." *Trends in Ecology and Evolution* 19 (12): 654–60.

Soron, Dennis. 2008. "Roadkill: Commodity Fetishism and Structural Violence." *Topia* 18:107–25.

Soulé, Michael, and Gary Lease, eds. 1995. *Reinventing Nature? Responses to Postmodern Deconstruction*. Washington, DC: Island Press.

Soulé, Michael, and Reed Noss. 1998. "Rewilding and Biodiversity: Complementary Goals for Continental Conservation." *Wild Earth*, Fall, 19–28.

Spash, Clive. 2015. "The Dying Planet Index: Life, Death, and Man's Domination of Nature." *Environmental Values* 24:17.

Speth, James Gustave. 2008. *The Bridge at the End of the World: Capitalism, the Environments, and Crossing from Crisis to Sustainability*. New Haven, CT: Yale University Press.

Steffen, Will, Paul Crutzen, and John McNeill. 2007. "The Anthropocene: Are Humans Now Overwhelming the Great Forces of Nature?" *Ambio* 36 (8): 614–21.

Steffen, Will, et al. 2011a. "The Anthropocene: Conceptual and Historical Perspectives." *Philosophical Transactions of the Royal Society A* 369:842–67.

Steffen, Will, et al. 2011b. "The Anthropocene: From Global Change to Planetary Stewardship." *Ambio* 40 (7): 739–61.

Steinberg, Ted. 2002. *Down to Earth: Nature's Role in American History.* Oxford: Oxford University Press.

Steiner, Gary. 2010. *Anthropocentrism and Its Discontents: The Moral Status of Animals in the History of Western Philosophy.* Pittsburgh: University of Pittsburgh.

Steiner, Gary. 2011. "Toward a Non-Anthropocentric Cosmopolitanism." In *Anthropocentrism: Humans, Animals, Environment*, edited by Rob R. Boddice, 81–115. Leiden: Brill.

Stephens, Tim. 2010. "Study Documents Widespread Extinction of Lizard Populations Due to Climate Change." http://news.ucsc.edu/2010/05/3787.html.

Stokstad, Erik. 2014. "The Empty Forest." *Science* 465 (6195): 397–99.

Stone, Dan 2013. "100 Million Sharks Killed Every Year, Study Shows on Eve of International Conference on Shark Protection." *National Geographic Magazine*, March 1. http://voices.nationalgeographic.com/2013/03/01/100-million-sharks-killed-every-year-study-shows-on-eve-of-international-conference-on-shark-protection/.

Strayer, David, and David Dudgeon. 2010. "Freshwater Biodiversity Conservation: Recent Progress and Future Challenges." *Journal of North American Benthological Society* 29 (1): 344–58.

Stromberg, Roland. 1975. *After Everything: Western Intellectual History since 1945.* New York: St. Martin's Press.

Suckling, Kieran. 2014. "Against the Anthropocene." *Immanence*, July 7. http://blog.uvm.edu/aivakhiv/2014/07/07/against-the-anthropocene/.

Sumaila, U. Rashid, and Daniel Pauly. 2011. "The 'March of Folly' in Global Fisheries." In *Shifting Baselines: The Past and Future of Ocean Fisheries*, edited by Jeremy Jackson, Karen Alexander, and Enrik Sala, 21–32. Washington, DC: Island Press.

Tal, Alon. 2017. "In a Crowded and More Violent Future." *Environmental Forum*, March/April, 51.

Taylor, Bron. 2000. "Bioregionalism: An Ethics of Loyalty to Place." *Landscape Journal* 19 (1): 50–72.

Terborgh, John. 1988. "The Big Things That Run the World—A Sequel to E. O. Wilson." *Conservation Biology* 2 (4): 402–3.

Terborgh, John. 1999. *Requiem for Nature.* Washington, DC: Island Press.

"Think Big." 2011. *Nature* 469 (January 13): 131.

Thomas, Keith. 1983. *Man and the Natural World: Changing Attitudes in England, 1500–1800.* Oxford: Oxford University Press.

Thoreau, Henry. 1991. *Walden, or, Life in the Woods.* New York: Vintage Books.

Thoreau, Henry. 1992. *Walking.* Bedford: Applewood Books.

Tilman, David, et al. 2001. "Forecasting Agriculturally Driven Global Environmental Change." *Science* 292 (5515): 281–84.

Tompkins, Kristine, and Tom Butler. 2016. "Expanding Parks." Editorial. *Outside*, October 10.

Trombulak, Stephen. 2011. "Becoming a Neighbor." In *The Way of Natural History*, edited by T. L. Fleischner, 126–36. San Antonio, TX: Trinity University Press.

Trombulak, Stephen, and Christopher Frissell. 2000. "Review of Ecological Effects of Roads on Terrestrial and Aquatic Communities." *Conservation Biology* 14 (1): 18–30.

Turner, Jack. 1996. *The Abstract Wild*. Tucson: University of Arizona Press.

Turvey, S. T., et al. 2010. "Rapidly Shifting Baselines in Yangtze Fishing Communities and Local Memory of Extinct Species." *Conservation Biology* 24 (3): 778–87.

Tuttle, Will. 2005. *The World Peace Diet: Eating for Spiritual Health and Social Harmony*. Brooklyn, NY: Lantern Books.

UNEP (United Nations Environmental Programme). 2012. "Growing Greenhouse Emissions Due to Meat Production." Global Environmental Alert Service. https://na.unep.net/geas/.

Urban, Mark. 2015. "Accelerating Extinction Risk from Climate Change." *Science* 348 (6234): 571–73.

Valliant, John. 2005. *The Golden Spruce: A True Story of Myth, Madness, and Greed*. New York: W. W. Norton.

van der Ree, Rodney, et al. 2011. "Effects of Roads and Traffic on Wildlife Populations and Landscape Function: Road Ecology Is Moving toward Larger Scales." *Ecology and Society* 16 (1): 48.

van Dooren, Thom. 2014. *Flight Ways: Life and Loss at the Edge of Extinction*. New York: Columbia University Press.

Vickers, Daniel, with Loren McClenachan. 2011. "History and Context: Reflections from Newfoundland." In *Shifting Baselines: The Past and Future of Ocean Fisheries*, edited by Jeremy Jackson, Karen Alexander, and Enrik Sala, 115–33. Washington, DC: Island Press.

Vidas, Davor. 2011. "The Anthropocene and the International Law of the Sea." *PNAS A* 369:909–25.

Vijay, Varsha, et al. 2016. "The Impacts of Oil Palm on Recent Deforestation and Biodiversity Loss." *PLOS One*, 1–19. DOI: 10.1371/journal.pone .0159668.

Vince, Gaia. 2011. "An Epoch Debate." *Science* 334 (October 7): 32–37.

Vörösmarty, C. J., et al. 2010. "Global Threats to Human Water Security and River Biodiversity." *Nature* 467:555–61.

Waldman, John. 2010. "The Natural World Vanishes: How Species Cease to Matter." *Yale Environment 360*, April 8.

Washington, Haydn. 2018. *A Sense of Wonder: Becoming Whole through Engaging with Nature*. London: Routledge.

Waters, Frank. 1972. *The Book of the Hopi*. New York: Penguin Books.

Weber, Max. (1905) 2013. *The Protestant Ethic and the Spirit of Capitalism*. Merchant Books.

Weeden, Don, and Charmayne Palomba. 2012. "The Post-Cairo Paradigm: Both Numbers and Women Matter." In *Life on the Brink: Environmentalists Confront Overpopulation*, edited by P. Cafaro and E. Crist, 255–73. Atlanta: University of Georgia Press.

Weis, Tony. 2013. "The Meat of the Global Food Crisis." *Journal of Peasant Studies* 40 (1): 65–85.

Weisman, Alan. 2007. *The World without Us*. New York: St. Martin's Press.

Weisman, Alan. 2013. *Countdown: Our Last, Best Hope for a Future on Earth?* New York: Little, Brown and Company.

Westbroek, Peter. 1991. *Life as a Geological Force: Dynamics of the Earth*. New York: W. W. Norton.

Whitehead, Alfred North. 1966. *Modes of Thought*. New York: Free Press.

Whitty, Julia. 2007. "Gone: Mass Extinction and the Hazards of Earth's Vanishing Biodiversity." *Mother Jones*, May/June, 36–45, 88–90.

Whitty, Julia. 2010. "The Last Taboo." *Mother Jones*, May/June.

Whitty, Julia. 2011. *Deep Blue Home: An Intimate Ecology of our Wild Ocean*. New York, NY: Mariner Books.

Wilcove, David. 2008a. *No Way Home: The Decline of the World's Great Animal Migrations*. Washington, DC: Island Press.

Wilcove, David. 2008b. "Animal Migration: An Endangered Phenomenon?" *Issues in Science and Technology* 24 (3).

Wilcove, David, and Martin Wikelski. 2008. "Going, Going, Gone: Is Animal Migration Disappearing?" *PLOS Biology* 6 (7): 1361–64.

Williams, Jerry. 2007. "Thinking as Natural: Another Look at Human Exemptionalism." *Human Ecology Review* 14 (2): 130–39.

Williams, Mark, et al. 2015. "The Anthropocene Biosphere." *Anthropocene Review*, 1–24.

Wilson, E. O. 1987. "The Little Things That Run the World." *Conservation Biology* 1 (4): 344–46.

Wilson, E. O. (1999) 2010. *The Diversity of Life*. Cambridge, MA: Harvard University Press.

Wilson, E. O. 2002. *The Future of Life*. New York: Alfred A. Knopf.

Wilson, E. O. 2006. *The Creation: An Appeal to Save Life on Earth*. New York: W. W. Norton.

Wilson, E. O. 2013. "Beware the Age of Loneliness." *Economist*, November 18, 143. http://www.economist.com/news/21589083-man-must-do-more-preserve-rest-life-earth-warns-edward-o-wilson-professor-emeritus.

Wilson, E. O. 2015. "Parks, Biodiversity, and Education." Keynote speech at Science for Parks, Parks for Science: The Next Century Conference, University of California, Berkeley, March 25.

Wilson, E. O. 2016. *Half Earth: Our Planet's Fight for Life*. New York: W. W. Norton.

Wittgenstein, Ludwig. (1953) 1968. *Philosophical Investigations*. 3rd ed. New York: Macmillan.

Wohl, Elle. 2013. "Wilderness Is Dead: Wither Critical Zone Studies and Geomorphology in the Anthropocene?" *Elsevier.* http://dx.doi.org/10.1016/j
.ancene.2013.03.001.

Wright, Ronald. 2005. *A Short History of Progress.* New York: Carroll & Graf.

Wroe, Stephen, et al. 2004. "Megafaunal Extinction in the Late Quaternary
and the Global Overkill Hypothesis." *Alcheringa: An Australasian Journal of
Palaeontology* 28 (1): 291–331.

Wu, J., and O. L. Loucks. 1995. "From Balance of Nature to Hierarchical Patch
Dynamics." *Quarterly Review of Biology* 70 (4): 439–66.

Wuerthner, George, Eileen Crist, and Tom Butler, eds. 2014. *Keeping the Wild:
Against the Domestication of Earth.* Washington, DC: Island Press.

Wuerthner, George, Eileen Crist, and Tom Butler, eds. 2015. *Protecting the Wild:
Parks and Wilderness, the Foundation for Conservation.* Washington, DC:
Island Press.

Zimmer, Carl. 2010. "A Looming Oxygen Crisis and Its Impact of the World's
Oceans." *Yale Environment 360,* August 5. http://e360.yale.edu/features/a
_looming_oxygen_crisis_and_its_impact_on_worlds_oceans.

Index